Alternate Energy:
Assessment and Implementation
Reference Book

Alternate Energy:
Assessment and Implementation
Reference Book

James J. Winebrake, Ph.D., Editor

THE FAIRMONT PRESS, INC.
Lilburn, Georgia

MARCEL DEKKER, INC.
New York and Basel

MARCEL

DEKKER

WITHDRAWN

COLORADO COLLEGE LIBRARY
COLORADO SPRINGS
COLORADO

Library of Congress Cataloging-in-Publication Data

Winebrake, James J.
    Alternate energy : assessment and implementation reference book /
James J. Winebrake, editor
        p. cm.
    I. Includes bibliographical references and index.
    ISBN 0-88173-436-5 (print) ISBN 0-88173-437-3 (electronic)
        1. Renewable energy sources--Handbooks, manuals, etc. I.
Winebrake, James J.

    TJ807.4.A48 2003
    333.79'4--dc22

                                                2003061562

*Alternate energy: assessment and implementation reference book / Winebrake,
James J.*
©2004 by The Fairmont Press, Inc. All rights reserved. No part of this
publication may be reproduced or transmitted in any form or by any
means, electronic or mechanical, including photocopy, recording, or
any information storage and retrieval system, without permission in
writing from the publisher.

Fairmont Press, Inc.
700 Indian Trail, Lilburn, GA 30047
tel: 770-925-9388; fax: 770-381-9865
http://www.fairmontpress.com

Distributed by Marcel Dekker, Inc.
270 Madison Avenue, New York, NY 10016
tel: 212-696-9000; fax: 212-685-4540
http://www.dekker.com

Printed in the United States of America

10 9 8 7 6 5 4 3 2 1

0-88173-436-5 (The Fairmont Press, Inc.)
0-8247-4289-3 (Marcel Dekker, Inc.)

While every effort is made to provide dependable information, the publisher, authors,
and editors cannot be held responsible for any errors or omissions.

Ref.
TJ
807.4
.A48
2004

for Susan

Table of Contents

Foreword

Over the past several years, industry and government have turned to a strategic planning technique called "roadmapping" to help understand and evaluate future energy management practices and technologies. In particular, "technology roadmaps" have been developed for a number of non-conventional energy technologies. These roadmaps identify technology trends, barriers to market entry, and technology and policy approaches for overcoming these barriers. Technology roadmaps represent an effective way to explore future technologies and to build relationships among important stakeholders who can move such technologies forward.

This book considers energy management and technology development over the next 20-25 years by exploring energy technology roadmaps. Drawing upon abridged and edited versions of roadmap reports, the book provides a unique multidisciplinary understanding of future energy technologies, their market potential, and the necessary policy actions required to achieve this potential.

With the exception of the first and last chapters, each chapter in the book presents a particular energy technology roadmap. The first chapter discusses roadmaps as a form of strategic planning and technology assessment. In this chapter, a general process for technology roadmap development is discussed. Chapters 2-9 present roadmaps of the following energy technologies:

- Chapter 2: Lighting Roadmap
- Chapter 3: Commercial Building Roadmap
- Chapter 4: Residential Building Roadmap
- Chapter 5: Windows Roadmap
- Chapter 6: Hydrogen Power Roadmap
- Chapter 7: Biofuels Roadmap
- Chapter 8: Wind Power Roadmap
- Chapter 9: Solar Power Roadmap

The final chapter (Chapter 10) pulls these roadmaps together to get a more holistic view of our energy future.

I am hopeful that many will find this book useful not only as a vision of sustainable energy futures, but also as a reference book for how technology roadmaps can be developed and used. I continue to encourage colleagues in industry and government to engage in technology roadmapping as a formal strategic planning activity. I hope this book will provide an impetus to others wishing to create their own roadmaps as a futures-thinking tool.

Acknowledgements

This book would not have been possible without the support and insights of F. William Payne, a friend and mentor. I have had many informative and often humorous discussions with Bill about energy technologies, government policy, and market realities. I also thank the folks at Fairmont Press for their patience and expertise in pulling together this volume. Finally, I thank my many colleagues in the energy field, particularly those at the U.S. Department of Energy, for their advice and support throughout the development of this book.

Note on photographs used in this book: Photos used in this book were obtained from the National Renewable Energy Laboratory's Digital Picture Library. These pictures did not appear in the original roadmap documents that make up the book's main chapters. Sources and credits for each photograph are included in the Appendix of this book.

Chapter 1

Introduction

WHAT IS AN ENERGY ROADMAP?

The term "roadmap" has come in vogue of late.[1] It has been used by politicians, engineers, statespeople, business leaders, and many others to describe a myriad of products and intellectual thought processes. Recently, the term "roadmap" has received significant attention in the press when U.S. Secretary of State Colin Powell introduced an Israeli-Palestinian Roadmap to Peace that articulates arrangements for a future peace in the Middle East.

From this book's interests the term roadmap is associated with something planners and futurists call "technology roadmaps." These roadmaps represent glimpses of the future—both the probable and the possible. Roadmaps are in many respects strategic plans that provide an organization (or collection of organizations) a template for outlining future goals and strategies to achieve those goals. In technology fields, roadmaps identify potential markets for new technologies and barriers to market development. Roadmaps also suggest a strategic approach that would lead to significant market penetration of these new technologies.

This book defines *technology roadmaps* as: "A futures-based, strategic planning device that outlines the goals, barriers, and strategies necessary for achieving a given vision of technological advancement and market penetration." Roadmaps provide us with a normative picture of the future (i.e., what we desire), an identification of barriers (what is preventing us from getting what we desire), and strategies for achieving that future (how we can overcome those barriers).

This book examines technology roadmaps that address sustainable energy technologies over the next 20 to 25 years. The book explores a set of roadmaps from a variety of industries, including:

- Lighting
- Commercial buildings

- Residential buildings
- Windows and Fenestration
- Hydrogen power
- Biofuels
- Wind energy
- Solar power

Exploring these technology roadmaps accomplishes two things. First, by examining the variant processes used to develop these roadmaps, we can identify some "best practices" for the roadmapping process. These techniques will help in the future development of technology roadmaps. Second, by understanding and synthesizing the trends, barriers, and strategies that each roadmap portrays, we can derive a more holistic view of the future of sustainable energy technologies.[2]

The remainder of this chapter discusses the roadmap processes used to create the roadmaps found throughout the rest of the book. Chapter 2 begins our collection of roadmaps from a variety of sustainable energy technology industries. The final chapter then synthesizes a more complete understanding of sustainable energy futures by calling upon content from the individual roadmaps and general energy trends.

SUSTAINABLE ENERGY ROADMAP DRIVERS

Over the past five years there has been a flurry of activity in the use of technology roadmaps in the energy field. In particular, roadmaps have emerged that explore the future of *sustainable energy technologies*. Although there is no hard-and-fast definition for sustainable energy technologies, one could modify the standard definition of sustainability[3] and suggest that sustainable energy technologies allow us to "meet today's energy needs without compromising the ability of future generations to meet their energy needs." Such technologies typically have the following attributes:

- They usually represent alternatives to conventional (i.e., fossil fuel) energy technologies;

- They are more efficient than current technologies, allowing greater energy services per energy unit input;

- They have a smaller impact on the environment than conventional energy technologies, for example by reducing local air pollutants, water pollution, or greenhouse gas emissions;

- They often employ renewable energy feedstocks, so that the resource constraints associated with conventional fossil fuels are addressed.

The two main drivers for sustainable energy "roadmapping" are: (1) concerns about environment quality, and (2) concerns about future energy supplies. Environmental concerns include air pollution at the local-level (e.g., tropospheric ozone, particulates, and air toxics), the regional-level (e.g., sulfur dioxide, mercury, particulates), and the global-level (e.g., climate change from greenhouse gas emissions). Conventional fossil fuel energy technologies are inherently carbon-based and combustion-based, and therefore environmentally problematic. And nuclear power, although not carbon-based, raises issues of nuclear waste management, nuclear material diversion for weapons production, and safety concerns that make it questionable as a long-term energy solution. Thus, sustainable energy technologies allow us to provide energy services today, without irreversibly harming the environment for future generations.

In the United States, the link between energy and environmental quality (particularly air quality) is stark. For example, fossil fuel combustion is responsible for approximately 80% of all greenhouse gas emissions in the United States, almost all the carbon monoxide and oxides of nitrogen emissions, about 60% of all emissions of volatile organic compounds, and about 90% of all sulfur dioxide emissions.[4] (These percentages are similar for other industrialized countries.) Thus, to address air pollution problems, the U.S. and other industrialized countries must directly target fossil fuel combustion.

The second driver towards sustainable energy roadmapping is energy supply. Energy supply represents the availability of secure, affordable fuel for long-term energy production. Because fossil fuels are nonrenewable, economic extraction of these fuels has a limited time horizon (albeit, the date of this time horizon is very much debated in the energy field). Conversely, energy from solar, wind, biomass, and water is renewable, and will theoretically only "run out" when the sun stops shining. Therefore, sustainable energy technologies allow us to supply our energy demands today without compromising the ability of future generations to meet their energy demands.

Sustainability of energy supply also takes on a political tone when we introduce the idea of energy security. Energy security can mean many things, but I argue that a reasonable definition is "the freedom from anxiety or fear that future quantities and prices of energy supply will be outside of acceptable ranges."[5] Although an energy resource may appear abundant globally, a given country may not have direct access to that energy resource; i.e., it must be purchased on world markets or through bilateral agreements. Because access to supply is out of the control of the dependent country, risk of long-term, adequate energy supply is increased.

Thus, from a national-interest perspective, sustainable energy technologies may also include technologies that use domestic feedstocks and resources. Such technologies would provide a sustainable supply of domestically-produced energy, thereby protecting a nation's energy supply against political or economic upheaval in other parts of the world. As the reader will see, this is an important attribute of "sustainability" in the roadmaps discussed in this book. These roadmaps have all been developed by entities in the United States; technologies that reduce the U.S. dependence on foreign sources of fuel are identified as "sustainable" and desirable by these organizations.

As an example of the potential concerns regarding energy supply, consider the U.S. transportation sector. In the U.S., 97% of the transportation sector is dependent on petroleum, and over 50% of that petroleum is imported at great economic and political cost. These trends in energy dependence are only forecast to get worse over the next 25 years.[6] Having a transportation sector that is almost entirely reliant on petroleum—over half of which is imported—has made the U.S. vulnerable to world oil market price fluctuations and supply disruptions. These disruptions can have significant impacts on the U.S. economy, and the specter of economic peril from constrained supplies can influence U.S. foreign policy in ways that are not in the long-term national interest.

Hence, sustainable energy technologies are identified as those technologies that can lead us to long-term, clean energy futures. The question that the roadmaps address is: What is the best way to advance research, development, and deployment of these technologies into energy markets? The sustainable energy technologies in this book face a myriad of technical, economic, and/or political barriers that must be overcome if they are to ever achieve significant market penetration. Roadmaps have emerged as one way for industry and government to collectively evaluate these barriers and identify strategies for overcoming them.

PURPOSE OF ROADMAPS

The roadmaps presented in this book have many purposes. These purposes are not mutually exclusive; indeed, in just about every roadmap you will see indications of each of the purposes highlighted below.

The overarching purpose of roadmaps is to *identify key goals, barriers, and strategies* for advancing a technology amidst technical, political, and market constraints and uncertainty. To accomplish this purpose, industry representatives, government officials, and academics meet to discuss the future outlook of a given technology, its potential, and the mechanisms for achieving market penetration. Identifying policy responses to overcome existing technical and market barriers is of critical importance in these roadmaps. These responses ultimately create the strategic plans that industry or government will pursue to prove and promote a chosen technology.

Another purpose of the roadmap effort is to *create dialogue* among industry, government, and other stakeholders. Such dialogue is of particular importance when the technology is not quite market-ready and public dollars are being directed for research, development, and demonstration (RD&D) of that technology. In these cases, industry needs to communicate with government to help guide the technology and policy agenda—and government needs to communicate with industry about the appropriate role of public support for RD&D. Because government incentives are often directed at these technologies, a public-private dialogue is required to assist in targeting those incentives properly. Too many times in the past government has developed incentives that have fallen short of expectations simply because their focus or design did not match industry or market realities. The roadmap process provides an opportunity for government and industry to exchange ideas and knowledge in a forum that is public and open.

Arguably, the most important outcome of the roadmap process is the *identification of strategies to overcome technical and market barriers*. All roadmaps discussed in this book identify barriers to market entry. Some roadmaps emphasize technical barriers—i.e., technical breakthroughs that will allow the technology to be more competitive with conventional energy technologies. Other roadmaps emphasize market barriers, such as overcoming consumer unfamiliarity with alternatives to conventional fuels. In almost all cases, an important outcome of the roadmap activity is suggested public policies to help the competitiveness of sustainable energy technologies.

THE ROADMAPPING PROCESS

There is no "official guidebook" for how to conduct energy roadmapping activities. There is no clearly defined process, and roadmaps have emerged from a many different procedural techniques. Sometimes these roadmaps begin with informal discussions (perhaps at an industry conference) that ultimately become a conference agenda item, and then a dedicated feature workshop. Other times, government agencies identify the roadmap as a necessary input in meeting its responsibilities, and so government coordinates a more formal process from the start.

Roadmaps often emerge from industry stakeholders, but because sustainable energy technologies typically require large-scale policy incentives, government has found a significant role in facilitating roadmap development. In the United States, the U.S. Department of Energy (DOE) has played an important role, sometimes leading the process of roadmap development.

Although often coordinated by government, roadmaps are conducted with input from a variety of stakeholders. In some cases discussed in this book, roadmaps have included over 200 industry representatives. In other cases, smaller stakeholder work groups and committees have met to develop roadmap content. Whatever the forum, because roadmaps address strategic planning issues over the next 20-25 years, all perspectives of the future are needed. Roadmaps tend to be richer when they consider a larger set of goals, barriers, and strategies from a variety of stakeholder inputs.

Specific roadmap development methodology is not a linear process;

there is much opportunity for feedback. In most of the cases explored in this book, stakeholders had opportunities to provide input into the roadmap process multiple times (examples are discussed in the appropriate chapters).

If one could generalize a roadmap process, a picture such as that in Figure 1-1 might emerge.

As shown in Figure 1-1, we can portray technology roadmapping as a six-step, interactive process. The first step is to *identify roadmap stakeholders and the coordinating body*. Sometimes, identification is an informal process—for example, emerging from discussions at an industry conference or government meeting. Other times, a coordinating body (such as the U.S. DOE) will begin the process and explicitly recruit stakeholders to participate.

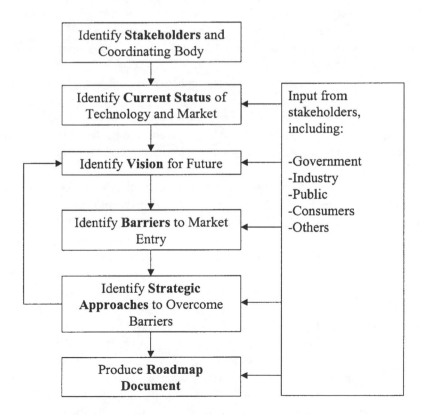

Figure 1-1. Generalized Roadmap Process

The second step is to *identify the current status of the technology and its market*. This is an important step in the process, for it requires that all stakeholders understand the technical starting point for the roadmap effort. In this step, engineers, scientists, and technologists present new technical advancements, and producers, distributors, and retailers share information to provide a clearer picture of the market.

The third step involves the *creation of a vision* for the industry. This step provides an opportunity for stakeholders to debate and discuss the market potential of the technology. This vision (often referred to as roadmap "goals") can cover various time frames. For example, roadmaps in this book include short-term (0-3 years), medium-term (3-10 years), and long-term (greater than 10 years) goals. In addition, some vision statements are short (one sentence) and make a very broad statement about the future of the technology; in other cases, the vision is segregated into several more specific statements that could include target values for market penetration or costs. Whatever the format, the stakeholder discussion that ensues during vision development is often the most interesting, contentious, and colorful of all discussions. Here we see input from stakeholders who span the spectrum of optimism for a given technology, and the coordinating body is often hard-pressed to help define a vision statement that is both positive, yet realistic. Anyone who has participated in this process can attest that accomplishing a common vision for these technologies is a difficult task.

Following the development of the vision, the fourth step of the roadmap *identifies the barriers* to the respective technology. These barriers usually include technical, market, and political barriers to market entry. Identifying barriers can be accomplished in a variety of ways. Usually, brainstorming with sub-groups of stakeholders is a useful method, followed by more sophisticated or structured methods of discussion or voting. Accurately identifying barriers is critical, as the strategic planning options that ultimately emerge from the roadmap are directly related to overcoming these barriers.

The fifth step of the roadmap is to *explore strategic planning* options to overcome the barriers from the previous step. A roadmapping method that has gained success in this area involves focus group discussions where the coordinating body or moderator steps through each barrier and participants brainstorm solutions to these barriers. This process is cycled through several times (e.g., in a Delphi method approach) until the list of strategic approaches becomes more focused. Often barriers and strategies are reorganized during this process— sometimes consolidated, other times separated—based on these discussions.

In addition, a feedback loop is identified from this step to the third step (*identify a vision*). This feedback represents the fact that sometimes thinking on strategic approaches elucidates necessary changes in the vision. In such cases, the vision is often revised, and new barriers and strategies are identified in relation to this new vision.

The final step is the *production of a roadmap document*. This book contains abridged versions of eight (8) roadmap documents produced by sustainable energy technology groups. How are these documents used once they are produced? There is a concern among many that such documents may simply "collect dust" on a bookshelf and never be used for proactive policy or technology development. Of course, different roadmap documents will find different uses, but at least three things could be said about these documents.

First, one valuable outcome of roadmapping emerges during the *process* of developing the roadmap document. In the process of developing the roadmaps, industry and government have an opportunity to share ideas, to network, and to obtain a better understanding of their individual perspectives on the technology of concern. The value of this network building is difficult to assess, but is surely positive.

Second, roadmap documents are used to help organizations (whether government agencies or industry groups) plan for the future. For example, government agencies have used roadmap documents as research planning tools for future RD&D expenditures. Industry groups have also used roadmap documents to provide a menu of policy incentives for educating consumers and lobbying lawmakers.

Third, roadmap documents provide an excellent overview on the current status of a given technology. It is rare to have a single document that reflects input from many stakeholders of a given industry regarding technical and market status (and barriers). For any researcher, analyst, consumer, or producer who is interested in a technology field, a roadmap document is a good place to turn for an introduction to the technology's status.

TECHNOLOGY ROADMAPS
USED IN THIS BOOK

The next eight chapters of this book present sustainable energy technology roadmaps developed for different industries. Each chapter represents and abridged and edited version of an actual roadmap developed within the last five years. The roadmaps include:

- *Chapter 2: Lighting Technology*. This roadmap examines the lighting industry and the opportunities for energy efficient and advanced lighting technologies over the next 20 years. This roadmap was coordinated by the U.S. Department of Energy.

- *Chapter 3: Commercial Buildings*. This roadmap explores technical advances that will make commercial buildings more energy efficient and more responsive to the needs of its occupants—particular through the use of a "whole building approach" to energy use. This roadmap was coordinated by the U.S. Department of Energy.

- *Chapter 4: Residential Buildings*. This roadmap discusses the future of building envelopes (primarily) in the residential building field. Barriers to technologies that provide higher efficiency and advanced energy management systems are explored. This roadmap was coordinated by the U.S. Department of Energy.

- *Chapter 5: Windows*. This roadmap explores the windows and fenestration industry and the technologies that will lead to smart, energy efficient, and aesthetically pleasing windows over the next 20 years. This roadmap was coordinated by the U.S. Department of Energy.

- *Chapter 6: Hydrogen Power*. This roadmap examines the "hydrogen economy" and the necessary technical advancements that are needed to increase the use of hydrogen as an energy source worldwide. This roadmap was coordinated by the U.S. Department of Energy.

- *Chapter 7: Biofuels*. This roadmap explores the opportunities for biomass to play a larger role in the U.S. energy landscape over the next

20 years. The roadmap looks at the technical and market barriers that must be overcome to increase the role of biofuels in the U.S. economy. This roadmap was coordinated by the Biomass R&D Technical Advisory Committee, administrated by the National Biomass Coordination Office and staffed by the U.S. Department of Energy and the U.S. Department of Agriculture.

- *Chapter 8: Small Wind Power*. This roadmap looks at the opportunities for small-scale wind power (turbines up to 100 kW in size). This roadmap was coordinated by the American Wind Energy Association, and represents one of two roadmaps in this collection that did not have a government agency or quasi-government agency as a coordinating body.

- *Chapter 9: Solar Energy from Photovoltaics*. This roadmap outlines strategies to promote solar power over the next 20-30 years. This roadmap was coordinated by the Solar Energy Industry Association, although it had significant participation from public and private organizations.

The final chapter of the book reviews these eight roadmaps in light of what they can tell us collectively about sustainable energy futures in the United States, and perhaps globally.

In abridging these reports, I have done my best to maintain the "flavor" of the roadmaps as originally produced. Although the roadmaps are organized slightly differently than in their original form, I believe that the focus of the reports remains true to the original intent. Of course, the reader interested in more detail is encouraged to review each of the roadmap documents in its original form.

References

[1] Industry and government have used "road map" and "roadmap" interchangeably. I use "roadmap" due to its pervasiveness in government documents that make up the bulk of this book.

[2] It should be noted that the roadmaps in this book were generated by U.S. interests; therefore, although international concerns are often considered, the content of the roadmaps is U.S.-focused. Despite this U.S. slant, the roadmaps still have much to offer international readers regarding energy technology status, market barriers, and strategic approaches, all of which

can be applied in an international context.

[3]Sustainability is usually understood to mean: Meeting the needs of the present generation, without compromising the ability of future generations to meet their needs. This definition is based on the 1987 Brundtland Report published by the United Nations.

[4]U.S. Environmental Protection Agency, *National Air Pollutant Emission Trends: 1900-1998* (March 2000) EPA 454/R-00-002.

[5]When formulating this definition, I called upon a number of illuminating discussions with Dr. Alex Farrell, University of California at Berkeley, to whom I am indebted.

[6]U.S. Energy Information Administration, *Annual Energy Outlook 2003*, Report #:DOE/EIA-0383(2003) Released January 9, 2003.

Chapter 2

Lighting Technology Roadmap

A Glimpse of the Future

By 2010... Lighting systems will have capabilities unlike anything before. Luminaires will become smarter and more integrated, communicating with control systems, performing self-diagnostics, and enabling preventive maintenance. New materials will make reflectors configurable and more integrated with the light source. Microelectronics will show up in smaller, more flexible ballasts, and sensors will provide multiple inputs to define the lighting environment for users. Controls will work with the larger building management system to optimize use of daylighting, thermal load management, preventive maintenance, and demand load shedding. Solid-state LEDs and organic light emitting polymers (LEPs) will become available in the market. Fluorescent sources will reach efficiencies approaching 200 lumens per watt while maintaining a high color rendition index (CRI) through the use of new two-photon phosphor coatings. Low-cost ballasts will increase flexibility of the systems, and low-cost electronic ballasts will make compact fluorescent lamps as common in America's homes as they are in the workplace.

By 2020... Design of building systems will optimally combine both natural and human-made lighting systems to shape the indoor climate. Technology will be available to capture daylight for later transmission and distribution. Programmable flat-panel luminaries will create theatrical effects that are currently unknown. The attributes of this light will be manipulated by advanced control systems. Highly efficient, reduced-mercury fluorescent sources will come to market, while incandescent lamps will see new life through advanced materials that will raise their efficiency to 60 lumens per watt.

INTRODUCTION

This chapter represents the combined thinking of the lighting industry, academia, and the research community to identify and chart the future course of the lighting industry. Importantly, this chapter outlines ways in which tomorrow's lighting demands can be met in a sustainable fashion. The "Lighting Technology Roadmap" presented in this chapter is also intended to provide guidance to the public and private sectors in planning future investments and initiatives.

This roadmap will help government align its research and development (R&D) activities and give industry associations a clearer picture of what technologies, policies, and market transformations are needed to achieve tomorrow's lighting goals. Like other Technology Roadmaps discussed in this book, this one was created through a constructive dialogue among government and industry stakeholders.

The next section of this chapter discusses the lighting roadmap development process. This is followed by a presentation of the lighting roadmap vision, including an overview of market and technology trends in the lighting industry. After the vision has been presented, the chapter identifies barriers to the vision (market-related and technological), followed by a section on strategies for overcoming those barriers. The chapter concludes with discussion of next steps for government and the lighting industry.

THE LIGHTING ROADMAP DEVELOPMENT PROCESS

The process of developing this lighting technology roadmap involved a series of three workshops among government and industry representatives. These workshops spanned a period of about a year.

The first workshop was held in September 1998, involving 20 executives from all sectors of the lighting industry. Participants were challenged with exploring the vast possibilities for improved lighting technologies, practices, and markets to 2020. The outcome of this workshop was a draft of the *lighting vision* statement and "big picture" goals of the industry.

The second workshop was held in December 1998 and included over 60 lighting industry stakeholders. At this workshop, participants identified potential market transformation and technology barriers to achieving the *lighting vision*. Participants also brainstormed ways to overcome those barriers.

The third workshop was held in July 1999 and included a dozen experts in the lighting industry. At this workshop, participants identified specific processes and technologies that extend from what is known today to the envisioned future of 2020. It was at this workshop that participants identified specific technical attributes and capabilities that are needed over the next 20 years for the *lighting vision* to be successful.

Finally, an internet "vote" took place involving 201 stakeholders in the lighting industry. Stakeholders voted on which of the market-transformation activities and future attributes of lighting technologies could have the greatest impact on achieving the *lighting vision*. These voting results ultimately formed the basis for the strategies and high-priority actions discussed in this Roadmap.

THE LIGHTING INDUSTRY VISION

The vision statement below was crafted by a group of government and industry experts in the lighting field. The developers of the vision statement explored market and technology trends and their implications for lighting. The vision statement reflects new opportunities for lighting based on these trends. This section of the roadmap provides information on each major trend.

VISION STATEMENT

In 2020, lighting systems in buildings and other applications will:
- Enhance the performance and well-being of people
- Adapt easily to the changing needs of any user
- Use all sources of light efficiently and effectively
- Function as true systems, fully integrated with other systems (rather than as collections of independent components)
- Create minimal impacts on the environment during their manufacturing, installation, maintenance, operations, and disposal

---

Marketplace Vision

The vision of the future marketplace for advanced lighting services is one of growth. There will be a growing demand for advanced lighting systems by businesses, building owners and managers, and end users.

In businesses, advanced lighting will support the relentless drive to increase productivity, create value-added services, and reduce costs. Tomorrow's lighting will respond to the significant changes now under way in the nature of work and, in turn, in commercial building design and management. For example, lighting of the future will enable more effective use of space for multitasking, so businesses can adapt workplaces currently designed for individualized, manual and paper-based operations into an environment that promotes teamwork, shared resources, and electronic processes.

Also in business, high-quality lighting systems increasingly will be valued for their ability to improve employee productivity, employee retention, and quality control, particularly as work becomes evermore dependent on information access and interconnectivity. Businesses and individuals also will gain greater understanding of how advanced lighting solutions can improve health, safety, and security in the workplace, as well as yield significant bottom-line savings by reducing energy consumption.

In the commercial sector, advances in lighting will help answer the needs of building owners and managers for the highest possible return on capital investments. Efficient, intelligent lighting systems—especially those networked in a "whole buildings" context with other building systems—will enable managers to exercise greater levels of control over building functions, minimizing operations, maintenance, and energy costs. More important, advanced lighting will provide the high-performance, aesthetically pleasing environment that increasingly will be demanded by tenants.

In the consumer market, advanced lighting will help fulfill our appetites for comfort, convenience, and instant information and connectivity. Whether at work or at play, as consumers we will demand an increased level of personalized control over the function and aesthetics of our lighting. Sensors and controls in future lighting systems will provide us with new levels of information about our environment, and will allow us to shape that environment to enhance our creativity and productivity.

Technology Vision

The lighting vision presented earlier anticipates a flood of scientific and technology developments that will make lighting an increasingly more effective, efficient, and dynamically responsive contributor to our built environments. Advanced lighting systems will exploit the capabilities of more powerful and cost-effective sensors and controls, wireless connectivity, high-efficiency light sources, breakthroughs in biotechnology and chemistry, innovative high-performance materials, sophisticated systems integrations and modeling capabilities, and many other new and emerging technologies.

Lighting design will increasingly be done in an integrated "whole buildings" context that optimizes human-made and natural systems (such as daylight) to provide efficient, high-quality lighting, heating,

Figure 2-1. White LED Technology

cooling, ventilation, and information exchange. Greater emphasis on ongoing education for lighting professionals, as well as outreach to consumers, will be needed to maximize the value and opportunities afforded by fast-changing technologies.

BARRIERS

As with many of the technology roadmaps presented in this book, there are a number of barriers that might prevent actualization of the vision. In this lighting roadmap, experts identified barriers to achieving the lighting vision. These barriers were categorized as "market-related barriers" and "technological barriers."

Market-Related Barriers

Market-related barriers represent challenges to transforming today's lighting markets to ones built upon demand for advanced lighting products and systems. A major factor threatening timely realization of the lighting vision is the traditionally low rate of technology develop-

Figure 2-2. Natural Lighting Example at Oberlin College

ment and product innovation in the lighting and building industries. This problem is highlighted in Figure 2-3.

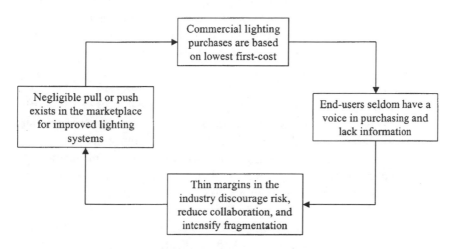

Figure 2-3. Cost v. Quality Cycle

In any industry, new products gain market acceptance over time by demonstrating a value superior to that of competing products. Yet product cycles have been exceptionally long in the lighting industry. The commercial building marketplace, in particular, has been slow to accept new lighting products and technologies, and the building industry has invested considerably less than most other industries in research, development, and demonstration. A study done in 1994, for example, found that U.S. private investment in construction research and development was only 0.5 percent of sales, while the rate of private investment for U.S. industry as a whole was 3.5 percent. One result of these long product cycles can be seen in the catalogs of today's lamp manufacturers. A typical manufacturer's catalog may carry 3,000 products, many of which have been available for decades.

Today there is negligible demand for innovation in commercial lighting, a situation that is worsened by the unusually complex distribution channel serving the construction industry. A general or electrical contractor on behalf of the building owner or manager often purchases lighting systems. While their purchases may be influenced by the recommendation of architects or light specifier, most contractors put an overriding focus on low first-cost lighting.

Indeed, contractors may often "value engineer" lighting, finding lower-cost options to substitute for the equipment selected by the architect or lighting specifier. The end user typically has little or no voice in lighting selection and often lacks awareness of the options available. In fact, an end user who seeks out information is likely to get conflicting recommendations from advisors in different parts of the distribution channel.

Because of the market focus on low-first-cost solutions, lighting equipment manufacturers are often pitted against each other to supply the least expensive system that will pass standards set by the installer, rather than the end user. As a result, lighting profit margins are increasingly constrained, limiting the industry's ability to invest in technology and product development. Further, since new products are accepted very slowly, production volumes of high-quality lighting solutions often remain below the critical mass needed to achieve economies of scale. As a result, high-quality lighting products stay pegged in high-priced niches, even though their market benefits and applications are potentially very broad.

In short, innovation in the commercial lighting marketplace is being neither strongly pulled by customer demand nor strongly pushed by industry investment. Transforming the dynamics of this marketplace— moving away from low-first-cost decisions to valuations based on life-cycle benefits—will be critical to achieving the lighting vision.

In addition, another market-related barrier to the lighting vision is the lack of a strong business case for advanced lighting that can drive end-user demand. Many case studies point to the advantages of high-quality lighting in improving productivity, employee retention, error reduction, and workplace safety; in attracting retail customers and improving retail sales; and in reducing energy consumption and other operation and maintenance costs. Yet these benefits have not yet been adequately documented, measured, and communicated to make a compelling case to tenants and building owners.

Tenants and building owners will be key in driving demand for higher-quality lighting, where life-cycle returns justify the greater initial cost of their purchase and installation. Architects, lighting specifiers, lighting manufacturers, and industry trade associations all will have pivotal roles in demonstrating and communicating these life-cycle benefits, and many will require ongoing education on advanced lighting technologies and design standards to perform these roles effectively.

Technological Barriers

Another set of challenges facing the lighting industry are technical, requiring increased levels of investment in science and technology, including basic and applied research, product development, and demonstrations. The industry participants who crafted the lighting vision identified several key technological trends that will affect commercial activities and buildings and, in turn, impose new demands on lighting technologies.

Commercial lighting systems will be challenged to adapt to the changing nature of work, including the increase in paperless, electronic-

Figure 2-4. Task Lighting Study at Lawrence Berkeley Lab

based tasks; the growing prevalence of team-based activities; and the ongoing reorganization and reconfiguration of many business function.

Another trend affecting lighting technologies will be the continued drive to enhance human productivity, creativity, and well-being. High-quality lighting that can measurably contribute to workplace productivity is expected to be in growing demand. In addition, there will be increased requirements for lighting systems to function as a fully integrated part of the total commercial building, raising the need for more sophisticated lighting control capabilities and for building design approaches that make optimal use of both natural and human—made systems.

A central challenge for future lighting systems is to allow end-user control of light intensity, color, color temperature, quality, and distribution within the user's space. Enabling control technologies need to be easy, intuitive, robust, and simple, and must be integrated with interoperable building-level controls.

In sum, lighting industry experts identified the following list as the most important technological barriers to overcome to achieve the lighting vision:

- Lack of standardization in current lighting control technologies and systems
- Need of more sophisticated control capabilities
- Need for better metrics to evaluate the quality and performance of products and technologies
- Lack of effective design tools and practices for integrating lighting into whole building design
- Need for higher-efficiency lighting sources, including reduced-mercury sources
- Need for new ballasts to support advanced lamp technologies
- Need of increased portability and flexibility in luminaries

STRATEGIES

To overcome market and technological barriers, the lighting roadmap identifies *strategies* necessary to transform the lighting marketplace. From a market perspective, strategic focus is on developing standards, encouraging consideration of life-cycle benefits, educating consumers, and providing incentives to reduce investment barriers.

From a technological perspective, strategic focus is on improving both the component parts of the lighting system (lamp, ballasts, controls, etc.) and the system as a whole (i.e., integrated into a larger building context).

Each of the strategies that emerged from discussions with experts is supported by a series of high-priority activities that will directly lead to their fulfillment. Virtually all of the activities and technology capabilities were judged to be achievable in the short (less than three years) to medium (three to 10 years) term. Some technology strategies are for a longer (more than 10 years) time horizon.

Market Transformation Strategies

Strategies related to market transformation are shown in the following tables. Each table highlights a main strategy and lists activities related to that strategy. The Time Frame column identifies whether these activities are in the short (S, 0-3 years), medium (M, 3-10 years), or long (L, greater than 10 years) term time horizon.

Table 2-1. Lighting Strategy #1.

Strategy #1: Develop clear definitions and standards for lighting quality.

Activity	Time Frame
Develop a uniform set of performance specifications for lighting systems.	S, M
Create industry-standard formats for energy and economics data used across the many available software packages.	M
Increase enforcement of ASHRAE/IESNA Standard 90.1-1989 and adoption of ASHRAE/IESNA Standard 90/1-1999.	S
Determine objective definitions and metrics of lighting quality.	S, M
Support and conduct third-party evaluation of integrated lighting system design and application.	S, M
Incorporate requirements for environmental sensitivity for lighting systems (such as daylight use) into the existing code structure.	M

Table 2-2. Lighting Strategy #2.

Strategy #2: Increase demand for high-quality lighting solutions by quantifying, demonstrating, and promoting life-cycle benefits to broad audiences.

Activity	Timeframe
Increase scientific knowledge of how lighting parameters impact human psychology, health, and productivity.	M
Maintain nonpartisan lighting centers and laboratories around the country where innovative lighting technologies can be demonstrated.	M
Conduct educational forums for end users about the effects of lighting on people and their activities.	S
Identify customer needs through needs assessments and focus groups.	S
Determine unique lighting characteristics and needs for specific environments (e.g., hospital, retail, office, restaurant).	M
Use work performance research to help make the marketing of quality lighting more effective, as with ergonomic furniture.	M
Conduct regional light fairs/expositions to showcase innovative lighting technologies.	S
Develop a marketing campaign promoting quality lighting to the general public.	S
Increase use of government buildings to demonstrate innovative technologies.	S, M
Increase publicity on the results of collaborative design and construction projects.	S, M

Table 2-3. Lighting Strategy #3.

Strategy #3: Strengthen industry education and credential lighting professionals.

Activity	Time Frame
Create educational programs on the design, installation, and use of lighting controls.	S
Improve education on daylighting, including simple rules of thumb for architects.	S
Provide training to product sales and distribution professionals about life-cycle cost analysis and the effects of lighting on people.	S
Increase information on NCQLP and CLMC certification requirements into Request for Proposal and Request for Quote language for building construction projects and energy-saving performance contracts.	S
Establish design assistance teams to teach others how to better integrate lighting into overall building design and how to maximize daylighting.	S, M
Increase use of the internet to provide information on research, demonstration, and regulatory activities.	S

Technology Development Strategies

Technology development strategies that identify the features and functions of tomorrow's lamps, ballasts, lighting controls, and overall systems are shown in the following tables. Once again, these tables identify the general strategy, activities related to that strategy, and the timeframe for those activities.

Table 2-4. Lighting Strategy #4.

Strategy #4: Accelerate the market penetration of advanced lighting technologies and system by providing incentives for R&D and reducing barriers inherent in today's specification and distribution methods.

Activity	Timeframe
Continue to develop rebate programs, coupled with public information programs, to transform the market for energy-efficient technologies.	S
Encourage manufacturers to develop new innovative and energy-efficient technologies through public/private programs.	S, M
Create new federal and state tax investment credits that encourage manufacturers and others to increase funding for lighting research.	M
Create supplemental construction project funding for the purchase of innovative lighting products (through government or manufacturer efforts).	S, M
Create longer-term funding commitments from Congress, rather than annual funding, for fundamental research programs.	M
Host stakeholder roundtables focusing on distribution channels and other issues affecting technology market penetration.	S, M

Table 2-5. Lighting Strategy #5.

Strategy #5: Develop advanced source and ballast technologies that enhance quality, efficiency, and cost effectiveness.

Activity	Timeframe
Achieve dimmability that maintains energy efficiency, color, and lamp life.	S, M

(Continued)

Table 2-5 (*Cont'd*)

Extend lamp life (less turnover).	M
Develop low-cost electronic ballasts for compact fluorescent lamps (CFLs).	S
Develop point source for optical fibers and pipes (high efficiency).	S, M
Create advanced solid-state structures such as LEDs, LEPs, and ceramics.	M
Maintain color throughout lamp life and from lamp to lamp.	M
Increase efficacy: greater than 100 lumens per watt at high CRI (*80 CRI).	M
For fluorescent lamps, develop two-photon phosphor technologies with efficiencies approaching 200 lumens per watt with CRI greater than 90.	M, L
For incandescent lamps, improve IR films to increase efficiency (50 to 100+ lumens per watt).	M
For incandescent lamps, improve efficiency of incandescing filaments by increasing the emissivity in the visible range (+10% to 15% efficacy) and increasing the temperature capabilities of these new materials (+25% to 30% efficacy).	M
For incandescent lamps, develop low-cost coatings to increase efficien-cies from the current level of 20 lumens per watt to 30 lumens per watt.	M
Develop improved design tools that incorporate daylighting concepts.	S, M
Develop toxic-free lamps and ballasts.	M

(*Continued*)

Table 2-5 (*Cont'd*)

Develop electrodeless metal halide technology, replacing mercury with xenon.	M
Develop new geometrical optics, efficient packaging, and efficient light distribution systems.	M
Create area sources (thin, flat panels).	M
Redesign ballasts and conduct materials research to solve the lumen depreciation/color shift problem that accompanies electrode degradation.	M
Develop new phosphor materials, electrode materials, and advanced ballast designs to produce gas discharges with quantum efficiencies greater than 1.5.	M
Develop universal ballasts. S, M	

Table 2-6. Lighting Strategy #6.

Strategy #6: Develop lighting controls with high levels of intelligence, interface capabilities, multiple levels of control, and ease of configuration.

Activity	Timeframe
Enable easy installation (e.g., self-configuring and friendly to non-experts).	S
Develop controls that are self-teaching, intuitive, easy to use.	S
Develop universal control and communication protocols for component interconnection.	M
Create a dialogue with energy management companies and lighting control industry in an effort to develop simple, easy-to-use controls. S, M	

(*Continued*)

Table 2-6 (*Cont'd*)

Incorporate anticipatory logic so systems learn and adapt to user preference.	M
Sense multiple inputs to configure and define lighting environments to user (color, room temperature, user temperature, user mood, eyesight of user, occupancy of room, motion, activity type, time of day, daylight levels).	M, L
Allow ease of programming by time of day and date.	S
Improve robustness (e.g., non-volatile memory)	S, M
Establish interactive linkage between the lighting, HVAC, and other system controls.	M
Provide some control at building level (range of levels, override).	S
Develop a universal building interface (remote control and monitoring) for load shedding, optimization of lighting/heat, preventive maintenance.	M
For public spaces, develop control systems that accommodate multiple uses of the space.	S
Develop control systems that serve emergency-response needs.	S
Develop control systems that monitor status of settings.	S

Table 2-7. Lighting Strategy #7.

Strategy #7: Develop luminaries and systems that enhance the quality and flexibility of light delivery.

Activity	*Timeframe*
Develop and utilize compatibility protocol to support "plug and play" (software and hardware).	M

(*Continued*)

Table 2-7 (*Cont'd*)

Utilize positioning and control to allow more effective task lighting.	S
Develop combined light source/reflector panel.	M
Achieve increased/variable reflectivity.	M
Develop materials that support multiple functions (e.g., reflect light and absorb sound).	M
Develop configurable reflectors.	M
Enable users to easily adjust quantity and direction of light from set location (e.g., adjustable louvers, configurable reflector/diffuser).	M
Develop smart fixtures that communicate with the control system, have intuitive learning capabilities, and perform diagnostics to enable preventive maintenance.	M
Develop expressive lighting that enhances psychological well-being.	S, M
Achieve foolproof installation and simplified operations.	S, M
Develop systems that capture daylight for later transmission and distribution.	M
Support easy movement of fixtures within a space.	S

CONCLUSION:
NEXT STEPS OF THE LIGHTING ROADMAP

This lighting roadmap outlines a view of where the lighting industry is today, a vision of where its stakeholders want to go tomorrow, and
strategies on how to get there; it provides guidance to both government
and industry on the direction of future activities; and it offers a frame-

work for greater collaboration across the industry in creating new market opportunities and innovative technologies, and provides guidance to government agencies in planning their activities and in forming R&D partnerships with industry.

The technology roadmap intentionally excludes detailed implementation approaches. These will be jointly developed between government and industry as the technology roadmap's strategies are analyzed and enriched. One early step in the implementation phase will be to investigate existing efforts already under way and determine how these might be leveraged to further the lighting vision and to avoid duplication of efforts.

For up-to-date information on implementation, refer to the lighting roadmap web site at *www.eren.doe.gov/buildings/research/roadmaps.cfm.*

[**Editor's Note:** This chapter represents an abridged and edited version of the U.S. Department of Energy report entitled *Vision 2020: The Lighting Technology Roadmap* published by the Office of Building Technologies, U.S. DOE, March 2000. The editor is grateful to the U.S. Department of Energy for permission to publish an abridged version of this report.]

APPENDIX 2-1: ACKNOWLEDGEMENTS FROM THE LIGHTING ROADMAP DOCUMENT

A NOTE FROM...

Dennis W. Clough, Team Leader, *Vision 2020: The Lighting Technology Roadmap*

As the Department of Energy's team leader for the development of *Vision 2020: The Lighting Technology Roadmap*, I have met and worked with literally hundreds of lighting professionals. Each of them has helped me better understand the many facets and intricacies of the lighting industry. These professionals have contributed to a process whose result will serve the government and industry for many years to come.

I would like to recognize three people especially:

Norm Grimshaw, Vice President of Technical Relations for Advance Transformer Company, spent many hours with me to ensure I understood the "real world" complexities of the industry. He has also been one of the lighting roadmap's most vocal and active advocates, consistently encouraging his colleagues and peers to engage in this important process.

Carol Jones, senior research scientist with Pacific Northwest National

Laboratory, introduced me to many of the industry's key players early in the roadmapping process, which helped make our executive forum, and ultimately the entire process, a success.

Ron Lewis, Director of Information Resources for Lighting Corporation of America and chair of NEMA's Lighting Systems Division, has been the model change agent for his industry. His forethought, insight, and integrative abilities moved this process to a level that would have been unattainable without him. Thank you so much for your help and guidance.

I would like to thank Battelle, Public Solutions, Inc., and What Box? Communications for their hard work with our executive forum and three roadmapping workshops, and thanks to Energetics, Inc., and Brandegee, Inc., for their support in the development, writing, and design of *Vision 2020: The Lighting Technology Roadmap*.

I would also like to thank the eight sponsoring professional associations, which helped us to get a wide representation of the entire industry engaged in this process. And finally, thank you to the 170 companies and organizations that actively participated in the development of the technology roadmap:

Available Light
Advance Transformer Company
Advanced Lighting Technologies
Alamo Lighting
Alliance to Save Energy
Armstrong World Industries
Auerbach & Glasgow
AVCA Corporation
Avista Utilities
Balzhiser and Hubbard Engineers
Battelle/Pacific Northwest National Laboratory
Belden, Inc.
Belfer Lighting Company
Benya Lighting Design
Bos Lighting Design
Building Acoustics and Lighting Labs
California Energy Commission
California Lighting Sales
CB Richard Ellis
CDAI Incorporated
CDS Associates, Inc.
CH2M Hill
Clark Engineers, Inc

Columbia Lighting
ComEd
Consolidated Edison of New York
Consultants in Lighting, Inc.
Cooper Lighting
Crownlite Mfg. Corp.
D & M Lighting Sales, Inc.
David T. Kinkaid Lighting Consultant
The DayLite Company
Devine Lighting
Duke Energy
Duke Solar
Dunham Associates
E. K. Fox and Associates
ECI Group, Inc.
Eclipse Technologies, Inc.
Electrical Design & Management
Electronic Lighting, Inc.
Energy Center of Wisconsin
Energy Controls and Concepts
Energy Savings, Inc.
Energy User News
Enron Energy Services
Environmental Systems Design, Inc.

EPRI Lighting Research Office
Eye Lighting International
Facilities Management
Field-Tech
Fitzgerald Lighting & Maintenance, Inc.
Florida A&M University
Fusion Lighting, Inc.
Gallegos Lighting Design
Gary Steffy Lighting Design, Inc.
General Electric Lighting
Genlyte Controls
Greg Fisher Lighting Sales
Grenald Waldron Associates
H.M. Brandston & Partners, Inc.
Hammel, Green and Abrahamson
Hayden McKay Lighting Design, Inc.
The HDMR Group
Hellmuth, Obata + Kassabaum, Inc.
HGA
Holophane
Honeywell- Home and Building Controls
Hubbell Lighting, Inc.
Illumination Concepts & Sales, Inc.
Indy Lighting, Inc.
Integrated Architecture
Irvine Engineering
JKAL
John Perry, AIA, LC
Johnson Controls Inc.
JR Frank Architect
Juno Lighting Inc.
Kansas State University
Keene-Widelite, Division of Canlyte Inc.
Kuyk & Associates, Inc.
Lam Partners Inc
Lawrence Berkeley National Laboratory
Ledalite Architectural Products
Light Bulb Supply Company, Inc.
Light Space, Division of Michaud
Cooley Erickson
Lighting Associates Inc
Lighting Corporation of America

Lighting Design Lab
Lighting Ideas, Inc.
Lighting Research Center, Rensselaer
Polytechnic Institute
Lighting Sciences Inc.
Lightly Expressed Ltd.
Lincoln Technical Services, Inc.
Lindsay-Pope-Brayfield
Lindsley Consultants Inc.
Litecontrol
Lithonia Lighting
LMS
Logan T. White Engineering
LUMEC Inc.
Luminae Souter Associates, LLC
Lutron Electronics
MagneTek, Inc.
Mars Electric Company, Inc
Matsushita Denko Shomei-Senryaku-
Kikaku
Matsushita Electric Works, Ltd.
MBA Consulting Engineers, Inc.
Michael John Smith Lighting Consultant
Mitchell B. Kohn Lighting Design
Moody Ravitz Hollingsworth
Mosher & Doran
Naomi Miller Lighting Design
National Air and Space Museum
National Council on Qualifications for
the Lighting Professions
National Research Council of Canada
NEES Companies
Nelson Electric, Inc.
Neoray Lighting
New Buildings Institute
National Renewable Energy Laboratory
New York State Energy R&D Association
Optika Lighting
OSRAM SYLVANIA
PAE Consulting Engineers
Philip Darrell Lighting Design
Philips Lighting Company

Power & Lighting Systems
Power Lighting Products / SLI Lighting
Precolite Moldcast
Reid Crowther & Partners Ltd
Richard Pearlson
Richard L. Bowen & Associates
RKL Sales Corporation
Robert Newell Lighting Design
Robson & Woese, Inc.
Roeder Design
RSA Engineering, Inc.
Ruda & Associates
San Diego Gas & Electric
San Jauquin Valley Electric League
Schneider Electric
Seattle Lighting Design Services
Selles Lighting Design
Sempra Energy Services
Shaper Lighting
Sylvan R. Shemitz Associates
Siemens Energy & Automation
Simkar Corporation

SLi Lighting
Spectrum Lighting Design
Steelcase
Studio Lux
Sunoptics
SW, Limited
TD Property Services
TEC Inc.
Tejas Potencial y Luminaria Asociados
TMT Associates
TransLight LLC
University of Nebraska
University Medical Center
Ushio America
Voss Lighting
The Watt Stopper, Inc.
Western Kentucky University
Winchester Electronics
Winona Lighting
Wright-Teeter Engineering Group
Zumtobel Staff Lighting, Inc.

Chapter 3

Commercial Buildings Technology Roadmap

A Glimpse of the Future

By 2020...Commercial buildings will feature:
- Organic, dynamic envelopes (like human skin) that can rapidly respond to changing environmental conditions
- Microscale thermal conditioning sources, individually controlled to meet user's preferences
- Dynamic, personalized ventilation
- "Plug-and-play" components and systems
- Waste source materials for building construction
- Solid-state sources for lighting, coupled with dynamic levels and daylighting
- Distributed energy resources at the site level (photovoltaic, fuel cells, combined cooling, heating, and power) generating power for the building or the utility grid
- Digital wireless microsensors, personalized building controls, and metering
- Buildings will be considered as part of a larger "whole community". The focus of building finance will become long-term, taking into account life-cycle benefits (versus today's 3-year horizon).

INTRODUCTION

Throughout much of human history, work and living spaces coexisted. Farmers lived on their land, merchants above their shops, craftspeople next to their forges and looms. The industrial revolution changed all that, as work became concentrated in factories and offices, at first in the vicinity of the labor force, and later, miles away along the trolley line or highway.

The separation between commercial and residential spheres grew ever sharper during the 20th century. Modern office buildings became possible with the advent of fluorescent lighting and air-conditioning, epitomized at mid-century by the sealed, self-contained International Style glass box. Today, our daily routines often take us from one specialized, comfort-controlled commercial facility to another—to work, learn, shop, and play—then home to distinctly residential communities.

Yet in the past few decades, some have begun to question how well commercial building technologies and practices serve emerging needs. Can we afford the environmental consequences of carrying the 20th century model into the future, or can we create commercial spaces that produce less waste, consume less energy, reduce reliance on cars, and minimize land use? Will specialized commercial facilities remain the norm, or will mixed-use buildings and communities better suit the way we live and work today? How must commercial buildings evolve to enhance human health and productivity, and to support the increasingly mobile, digital, and team-based nature of today's businesses?

This commercial building roadmap represents the results of a two-year process aimed at understanding these and many other questions. The roadmap evaluates the promise of new technologies and practices in the design, planning, siting, construction, and operation of commercial buildings. This chapter describes the commercial building vision and strategies for creating a commercial building landscape that meets future comfort and productivity needs.

THE COMMERCIAL BUILDING
TECHNOLOGY ROADMAP PROCESS

The federal government was the organization that fostered the development of this technology roadmap. Given that the federal government is the largest owner and operator of commercial facilities in the nation, it had a strong incentive to bring public and private stakeholders together to address commercial building concerns.

To produce the roadmap, the U.S. Department of Energy (DOE) developed a series of five workshops aimed at exploring the future of commercial buildings. In these workshops, participants—including architects, engineers, lighting and other designers, equipment manufacturers, researchers, building owners and developers, facility managers, building trades representatives, utility and energy service company representatives, and financiers—discussed the current state of the industry, significant trends and opportunities, and ways to align public and private research and development with real-world needs. They also identified areas of market transformation and education where industry participants could cooperate and where the federal government could play an expanded role. In all, more than 250 individuals from 150 different organizations participated in this technology roadmap development. A description of the workshops follows:

- The first workshop was held on July 27, 1998. There, 36 representatives of the building industry, associations, and government met to develop a vision statement and strategic goals for commercial buildings in the year 2020.

- The second workshop, in October 1998, included 66 designers, developers, and representatives from the building trades. These participants examined the forces driving or impeding whole-buildings approaches to design, siting, construction, and commissioning.

- The third workshop was held in January 1999 and included 76 representatives of various organizations who met to define principal gaps and needs in technology and processes related to the operation and maintenance of commercial buildings of the future. These participants created detailed action plans to meet identified strategic needs.

- The fourth workshop was held in April 1999 and included nine futurists and visionaries. These futurists developed a vision of the technology of the built environment in *2050*.

- The fifth and final workshop was held in October 1999. Here, 79 representatives of the building industry, associations, and government reconvened to develop a prioritized list of activities to further the whole-buildings approach within the commercial building sector.

THE COMMERCIAL BUILDINGS VISION

The Shape of Commercial Buildings in 2020

Major social, economic, technological, and environmental trends are changing the way we work, learn, and play. These changes, in turn, will create new demands on commercial buildings of the future. Here are some of the most evident trends and their possible implications in the coming decades.

Knowledge-based work. With the ongoing growth of the information-based economy, people will be increasingly engaged in highly visual and analytical work. Commercial buildings will be expected to provide reliable, continual, and instantaneous connectivity to information and electronic communications resources. Information technologies will no longer be captive in desktop computers, but will be distributed within the commercial environment, integrated into everything from furniture to windows. Demand will grow for personalized control of lighting, temperature, ventilation, and other aspects of the interior environment to enhance the productivity of knowledge workers.

Collaborative, reconfigurable workplace. Advanced communications and computing technologies will enable coworkers to collaborate ever more effectively from remote locations, decreasing the need to spend the workweek in shared physical spaces. When colleagues do work together, they will more often require flexible and reconfigurable space to accommodate term-based activities and frequent organizational and operational shifts. Education will also become more reliant on electronic technologies and team-based activities, redefining the requirements for future schools, libraries, and other learning facilities.

An aging, shifting population base. The mean age of the U.S. population continues to trend upward, increasing the need for ease of access and mobility within commercial facilities. Population will continue to increase in our deserts and on our seacoasts, two fragile ecosystems, requiring increased attention to resource efficiency, energy efficiency, and sustainable practices in commercial buildings.

Urban rebirth. Another trend having an effect on commercial building is the rebirth of urban centers and the corresponding need to reconfigure existing buildings for new uses. To stem over-development in suburban areas, increasing numbers of communities will enact zoning and create incentives to encourage the movement of businesses and residences back into the city.

Construction labor shortages. Demographic and economic shifts will continue to reduce the pool of skilled construction workers, necessitating less labor-intensive building methods and technologies in the future.

Environmental and health issues. Increasing public concern about environmental issues, coupled with the potential for more stringent environmental regulation, will drive market demand for commercial buildings that minimize resource use and waste in their construction and operation. Demand for healthier and more comfortable indoor environments will also grow as environmental awareness encompasses indoor as well as outdoor areas.

Energy issues. Greater cost-competitiveness of photovoltaics, fuel cells, and combined heat and power—coupled with the purchasing flexibility created by utility restructuring—will make on-site power generation an increasingly viable option for commercial buildings, where energy is becoming increasingly important (see text box below). Shrinking capacity margins in baseload power generation, and resulting concerns about the reliability of power, will further fuel this trend. Demand will also grow for energy-efficient buildings, particularly in areas with relatively high power costs or reliability concerns. Any future controls on carbon dioxide emissions will accelerate the demand for "green" power, renewable energy, and energy efficiency.

Insurance and liability issues. Insurers will exert pressure on the commercial building industry to increase the safety and longevity of buildings. Insurance will be increasingly expensive or unavailable for buildings constructed "in harm's way," e.g., in flood plains or seismic hot spots. Insurers will increase their involvement in building code development and enforcement. Both builders and building component manufacturers will be subject to higher liabilities for failures. Lawsuits related to indoor air quality and other health issues will become more prevalent.

Creating the Vision

In developing this technology roadmap, stakeholders developed their vision for the year 2020, shown in the text box below. Stakeholders also identified barriers to be addressed and defined strategies that will help make their vision a reality. Each strategy, as well as associated activities and milestones, is discussed in this chapter.

The Energy Dimension of Commercial Buildings

Today, the 4.6 million commercial buildings in the United States account for approximately one-sixth of total national energy consumption, or 16 quadrillion Btu and 32 percent of total national electricity consumption. Consumption of electricity in the commercial buildings sector has doubled in the last 18 years, and can be expected to increase by another 25 percent by 2030 if current growth rates continue.

Making commercial buildings more energy- and resource-efficient represents an enormous opportunity to save money and reduce pollution in every community across the country. Indeed, with annual energy expenditures in the commercial buildings sector of $100 billion, an efficiency improvement of 30 percent would yield $30 billion per year in bottom-line savings.

Benefits to the environment would also be substantial, including reduced emissions of sulfur dioxide, nitrogen oxides, and carbon dioxide from fossil-fueled power generation.

Such a 30 percent improvement in energy efficiency can be realistically achieved in the coming decades by applying existing technologies. Even more dramatic improvements—ranging from 50 to 80 percent—could be achieved with aggressive implementation of this technology roadmap, including a long-term approach to research and development.

Ultimately, the appropriate use of combined heating, cooling, and power systems, optimized building controls, solar and other forms of renewable energy, and energy-efficient building shells and equipment can produce commercial buildings that become net electricity generators rather than consumers.

Looking Forward

How might these "healthy, productive, and desirable" commercial buildings look and perform? Tomorrow's high performance commercial buildings are likely to:

Incorporate smart, responsive technologies. Commercial buildings will become almost "alive," using smart materials and systems that sense internal and external environments, anticipate changes, and respond

---

Vision Statement

By the year 2020:

- Successful public/private partnerships will deliver highly adaptable, sustainable, cost-effective commercial buildings.
- Advances in building design and operation will provide simple solutions to address the complex interactions of systems and equipment.
- Occupants, owners, builders, and communities will value America's commercial buildings as healthy, productive and desirable places to learn, work, and play.

---

dynamically through a "whole-buildings" approach (see text box below). Through wireless sensors and controls, energy using components will monitor when and how much they are needed and will adjust their operation accordingly. Individualized control of lighting, ventilation, and thermal conditioning will become possible, and user profiles that specify personal environmental preferences will follow an individual through a building or group of buildings. Uniform protocols will allow control devices to talk to each other and communicate externally. Buildings will aggregate performance information, self-diagnose and correct problems, and alert users to causes of substandard operation.

Reflect sound environmental practices. Tomorrow's commercial buildings will be highly resource-efficient and will make use of environmentally sustainable (low embodied energy) materials. They will also operate efficiently, using 30-80% less energy than 20th century buildings. Some will even be net electricity exporters, generating their own power through such on-site technologies as fuel cells and photovoltaics, and supplying excess power back to the grid. Sunlight will be used increasingly to produce electricity as well as for daylighting. Passive solar construction and natural ventilation will be regularly incorporated. Buildings will be designed for much greater flexibility and adaptability to reuse, resulting in longer life. Components and materials will also be designed for complete recyclability at the end of their lifetimes.

Be an integral part of sustainable community development. Commercial buildings will become more closely integrated with the surrounding environment. Building philosophy will shift from design of

---

What is the "Whole Buildings" Approach?

Today's commercial buildings employ complex and diverse technologies in their construction, operation, and maintenance. Building materials, components, and subsystems traditionally have been designed and implemented based on standardized criteria that are largely independent of one another.

For example, water-heating loads are considered to be solely a function of building use and are calculated independently of a building's plumbing design. Potential interactions between the two functions—for example, heat recovery from outgoing wastewater for pre-heating the incoming supply—are usually ignored.

Through a whole-buildings approach—sometimes referred to as "systems engineering"—all of the building components and subsystems are considered together, along with their potential interactions and impact on occupants, to achieve synergies. The fundamental goal is to optimize the building's performance—in terms of comfort, functionality, energy efficiency, resource efficiency, economic return, and life-cycle value. The whole-buildings approach crosses disciplines—requiring the integration of planning, siting, design, equipment and material selection, financing, construction, commissioning, and long-term operation and maintenance.

Implementing a whole-buildings approach has been shown to enhance air quality, lighting, and other key aspects of the building indoor environment. The natural environment benefits as well—through energy and waste reduction and more effective land use.

---

single, stand-alone buildings to campuses or even communities. Resource management will be optimized across the entire community, through strategies such as distributed power generation. More building space will perform double duty as both commercial and residential space. Fewer but better buildings will be constructed as a consequence. Communities will benefit from better land and resource use, better quality of life, and lower investments in highways and transit, and will structure tax and zoning policies to encourage whole-building development.

Be recognized for their bottom-line benefits to businesses and de-

Figure 3-1. Zion National Park Visitor Center and Passive Down-draft Cool Tower

velopers. By enhancing occupant productivity, health, and safety—and reducing life-cycle energy and operating costs—high-performance commercial buildings will make measurable contributions to the bottom line of tenant businesses. Financiers and insurers will acknowledge high performance buildings through favorable lending and underwriting practices and will also market high-performance building modifications as an option to their customers. Developers will realize better asset value as a result of the strong market appeal, adaptability, and long life of high-performance commercial buildings.

   Be designed for simplicity and safety. Future buildings will be ever simpler to construct and operate. Design and building techniques will enhance construction safety, reduce development and construction time, and cut labor intensity. Building controls and subsystems will be intuitive and elegant, requiring minimal technical expertise to operate and maintain.

CASE STUDY EXAMPLE:
A MODEL IN MANHATTAN

Project: 4 Times Square
Developer: Durst Organization
Project Architect: Fox & Fowle Architects, P.C.
Construction Manager: Tishman Construction Corporation
Project Engineer: Cosentini Associates

4 Times Square, at the intersection of Broadway and 42nd Street, is the first Manhattan office tower to incorporate "green" standards—energy-efficient design, indoor ecology, sustainable materials, and on-site power generation. Highlights of this 1.6 million-square-foot, 48-story building include:

- The ability to generate some of its electricity with on-site fuel cells—large, natural gas "batteries" that create power through a chemical reaction. They run cleanly and quietly 24 hours a day. No combustion is involved and waste products are hot water and CO_2.

- The use of building-integrated photovoltaic (PV) panels on limited areas of the facade. Peak output is about 15kW, enough electricity to run five suburban homes.

- The use of DOE-2, state-of-the-art software for analyzing a building's energy use. It can accurately model and compare potential energy savings from a variety of technical options.

- A ventilation system that provides tenants with 50 percent more fresh air than required by code.

Use of whole-building standards has reduced energy costs at 4 Times Square by an estimated $500,000 annually compared to expected costs in a traditionally constructed building, resulting in a payback period of five years or less.

BARRIERS

Need for Innovation

By 2020, high-performance commercial buildings can be making substantial contributions to sustainable community development and environmental protection, and also returning healthy bottom-line benefits to tenant businesses in the form of energy savings, operational savings, and productivity improvements. But what will it take to get there?

Overcoming technology barriers will certainly be vital. Achieving the integrated, "smart" buildings of the future, together with higher levels of energy- and resource-efficiency, will require continued research and development, with a focus on system integration and monitoring, as well as component optimization.

Although technology challenges are significant, they are dwarfed by the need for:

- Clear performance metrics that make a compelling economic case high-performance commercial buildings;
- Process changes by which building planning, design, construction, and operation and maintenance are conducted; and,
- Market transformation to overcome the current lack of demand for high-performance commercial buildings.

Each of these barriers, including the technology challenges, is discussed in the sections that follow.

Need for Clear Performance Metrics

A compelling case for high performance commercial buildings must be proven. The whole-buildings approach seeks to achieve low total costs over the life of the building, by minimizing energy and resource consumption, simplifying operational and maintenance requirements, and extending building life. However, the first costs for a whole buildings development can often be higher than for traditional approaches. For example, the whole-buildings approach may entail higher capital expenditures for sophisticated lighting and windows, nontraditional construction to maximize daylighting, and investments in on-site power generation equipment. It may also require additional up-front investments by the owners, developers, designers, contractors, and other key parties.

Justifying higher initial costs is a difficult sell in the commercial world, where the driving force is to keep first costs as low as possible. Life-cycle operational costs and performance issues—such as the quality of the indoor air or functionality of lighting—are seldom on the table at contract-signing time. Developers and builders generally have no stake in the long-term operating costs or performance of the building, and are rewarded based on their ability to control first costs. The ultimate building occupants typically have little voice in design and construction decisions, and are seldom able to quantify how the benefits of lower operational costs or improved building performance might justify a higher initial investment.

Measurable, defensible, and reproducible financial returns will be needed to create markets for commercial whole buildings. What are the broad-based, life-cycle benefits—in energy and resource use, operational cost savings, asset value, productivity of tenant businesses, and sustainable community development? And what returns will developers and communities realize by investing in high-performance buildings? Anecdotal evidence, while valuable, is not sufficient to spark widespread adoptions of whole-buildings approaches, particularly given the large investments and risks involved in a typical commercial building. Determining which performance metrics are of greatest value and their most reliable means of measurement and reporting are core challenges.

Technology Challenges

Research, development, and deployment efforts will be essential in realizing the vision for high-performance commercial buildings. Improvements are required in building components and equipment, as well as in how these elements are integrated within a whole-building, and even whole-community, systems context.

Systems integration challenges—rather than component-level technologies—are the focus of this technology roadmap. Despite tremendous advances in computing and control technologies, most buildings are still relatively "dumb" in their operation. For example, even though heat generated by lighting and office equipment is integrally related to building heating and cooling loads, these building functions are generally operated independently and are often at odds. Interior offices are cut off from natural light. Ventilation systems are out of sync with the configuration of offices and hallways. Buildings respond reactively to external conditions rather than proactively anticipating them, and inefficiencies abound. The results are utility bills higher than necessary and tenant dissatisfaction.

The commercial buildings vision calls for new technologies to overcome the inefficiencies. In particular, it foresees whole building design tools and smart, integrated building controls that enable optimized interactions among such subsystems as heating, lighting and daylighting, ventilation, and the building envelope. Whole buildings will be designed for optimal utility, environmental performance, and life-cycle value, and will essentially control themselves to maintain targeted performance. Critical barriers that must be addressed include lack of standard protocols for interoperability, difficulty and expense in retrofitting existing buildings, and restrictive building codes. In addition, advances will be needed in the performance and cost of sensors and wireless control technologies.

Figure 3-2. Intelligent Building Design at the National Renewable Energy Lab

Requirements for Process Changes

Developing a high-performance commercial building is a team effort, requiring close collaboration among building owners, architects, engineers, financiers, managers and operators, building trades representative, contractors, and other key players. Typically, this collaboration begins by reaching agreement on the system requirements—i.e., the performance targets (economics, energy, productivity, resource, recyclability) to be set for the building. In order to make the trade-offs among these building attributes, collaboration is necessary throughout the siting, design, construction, and commissioning. In the years to follow, ongoing collaboration is also needed to monitor building performance and evaluate lessons learned.

This kind of integrated building design and construction process departs radically from the approach used today, in which each discipline in the fragmented development process performs its work largely in isolation from the other and often with very different driving goals. Its widespread adoption will require new channels, tools, and methodologies for collaborative communication, problem solving, and decision making across these disciplines. Restructuring compensation and incentives may also be necessary—for example, basing a portion of fees, commissions, and rental incomes on how successfully the building achieves targeted performance requirements.

Market Transformation Challenges

Stimulating market demand for high-performance commercial buildings hinges on demonstrating a compelling economic case. In addition, several critical barriers must be overcome:

1. *Fragmentation within the commercial buildings "industry."* A defining characteristic of the U.S. building sector is its fragmentation. Hundreds of thousands of companies of all sizes design, build, finance, equip, own, or manage commercial buildings. Collaboration and communication among these companies is minimal. This high degree of fragmentation greatly complicates the process of implementing and marketing commercial whole-building concepts, since no single company or professional association influences the full range of disciplines and functions involved. It also limits private sector research, development and deployment of new technologies. Individual companies are seldom large enough to risk sizable investments on their own or to capitalize on any resulting innovation.

2.   *Financing barriers and tax disincentives.* Commercial building is by nature speculative and uncertain. Financing tends to reward conservative practices and impede innovation. When tax incentives and special financing options are offered, they are usually for individual components (such as HVAC systems) rather than for whole-building design approaches. As a result, developers and designers first pursue options for which incentives are offered, and design strategies for whole-buildings may be sub-optimal or neglected altogether. In addition, the current tax code actually discourages saving energy; energy costs are deductible against income—thus saving energy may actually increase a building owner's income tax liability.

3.   *Lack of holistic regional planning.* Commercial building locations are still determined based on an underlying assumption of cheap transportation and continued road building. As a result, many communities are marked by a sharp distinction between where people work and where they live. Given the long lifetime of such infrastructure, regions can be locked into inefficient patterns for decades. In contrast, a commercial whole-buildings approach to urban planning and site development might engage regional decision—makers in evaluating the cost savings and environmental benefits of building fewer roads and reducing commuter traffic, and might weigh these factors in a total cost/benefit evaluation. Effective models of holistic regional planning will be required, together with metrics that demonstrate the financial returns to communities.

STRATEGIES

Four interrelated strategies will be key to advancing the high-performance commercial buildings vision. Each strategy must address the unique requirements of rehabilitation projects, as well as of new construction. These strategies are tied to each of the barriers discussed above.

Strategy #1: Performance Metrics

Establish core definitions and metrics for high-performance commercial buildings. To do this, we need to:

- Define what to measure—i.e., determine the central characteristics of high-performance commercial buildings.
- Define how to measure—i.e., determine methods for measuring performance of commercial whole-buildings over time (building performance indices). This may include national protocols for organizing, storing, and retrieving this information.
- Determine how to apply the metrics to enable key audiences to evaluate costs and benefits of high-performance building investments—e.g., establish methods for evaluating total life-cycle costs and benefits for owners, occupants, and communities.

Strategy #2: Technology Development

Develop systems integration and monitoring technology that enable whole buildings to achieve optimal, targeted performance over their life cycles. To do this we need to:

- Develop verifiable design and performance analysis models and tools that enable component and system optimization (e.g., automated decision support tools)
- Develop methods to improve interoperability among architectural, mechanical, electrical, plumbing, and other key building subsystems, working with standards organizations.
- Develop cost-effective, reliable monitoring and control technologies (e.g., indoor air quality sensors, wireless sensors and controls) to ensure that performance targets are met throughout building life.

Strategy #3: Process Change

Create models of collaborative high-performance commercial building design and development, and establish the tools and professional education programs need to support these processes. To do this we need to:

- Develop, pilot, and document new models of collaborative whole-building design and development, and create implementation guidelines for applying such processes.
- Create tools (e.g., software, communications) to support integrated decision-making in commercial building design, construction, operation, and renovation.
- Establish education programs for professional who are key to implementing and supporting commercial whole-buildings approaches. For example, the development of a whole-building cur-

ricula as an integral part of formal education and continuing education for architects, designers, and engineers.

Strategy #4: Market Demand
Stimulate market demand for high-performance commercial buildings by demonstrating and communicating compelling economic advantages. To do this, we need to:
- Demonstrate and document the economic case for high-performance commercial buildings through pilots and case studies.
- Define and promote tax and financing incentives that would support commercial whole-building approaches, including the pursuit of tax incentives, financial market discounts, favorable insurance policies and federal subsidies for research.
- Develop and implement a strategic communications and marketing plan addressing all key audiences. This would involve communicating successes and best practices and sponsoring competitions and conferences to showcase new best practices.
- Develop and promote a "brand name" and identify for high-performance buildings (e.g., a simple well communicated program similar to the Energy Star model).

NEXT STEPS

This chapter outlines an ambitious vision for the commercial buildings industry. It serves as a resource for the public and private sectors and offers a framework for greater collaboration across the industry in creating new market opportunities for high-performance commercial buildings. The roadmap is also useful in providing guidance for future private and public research and development focus.

The roadmap intentionally excludes detailed implementation approaches. Such approaches will be jointly developed over time between government and industry. For readers interested in following the process of this particular roadmap, please go to: *http://www.eere.energy.gov/buildings/research/roadmaps.cfm*.

[**Editor's Note**: This chapter represents an abridged and edited version of the U.S. Department of Energy report entitled *High-Performance Commercial Buildings: A Technology Roadmap* published by the Office of Building Technologies, U.S.

DOE, March 2000. The editor is grateful to the U.S. Department of Energy for permission to publish an abridged version of this report.]

APPENDIX 3-1:
ACKNOWLEDGEMENTS FROM THE
COMMERCIAL BUILDINGS ROADMAP DOCUMENT

A NOTE FROM...

Drury Crawley, Team Leader, *High-Performance Commercial Buildings: A Technology Roadmap*

As team leader of the roadmap development process for DOE's Office of Building Technology, State and Community Programs (BTS), I wish to thank the hundreds of organizations and individuals that contributed to this significant effort. In particular, I want to acknowledge the crucial guidance of the Commercial Roadmap advisory group: Bill Browning, Rocky Mountain Institute; Jim Cole, California Institute for Energy Efficiency; Rick Fedrizzi, Carrier Corporation; Jim Hill, National Institute of Standards and Technology; Steve Kendall, Housing Futures Institute, Ball State University; Gail Lindsey, Design Harmony; Tom Phoenix, Moser Mayer Phoenix Associates; and Jim Yi, Johnson Controls.

Without Sean McDonald and Bruce Kinzey of Pacific Northwest National Laboratory, this roadmap could not have happened—they organized the workshops (and the participants), documented the workshops, and summarized material—creating the first drafts of this roadmap.

I would also like to thank Doug Brookman, Public Solutions, for his expert creative facilitation for the four workshops—crafting meaning and structure out of the chaos. Finally, I want to thank Karen Marchese, Nancy Reese, Karen Snyder, and Julie Tabaka, of Brandegee, for their vision and creativity in bringing the roadmap together into the document you see today.

Throughout the development of the roadmap, the participants have been an extraordinarily inspired, energetic, and expert group. We look forward to working with them and many others in realizing their vision for high-performance commercial buildings.

The organizations that participated in developing this roadmap include:

AEP
AFL-CIO
Air-Conditioning and Refrigeration Institute
Alfred University
Altieri Sebor Weiber Engineers
American Express Company

American Gas Cooling Center
The American Institute of Architects
American Iron and Steel Institute
American Society of Heating, Refrigerating and Air-Conditioning Engineers
Antares Group
Armstrong World Industries

Arthur D. Little
Ball State University
Barry Donaldson & Associates
Bevilacqua-Knight
Brandegee
British Columbia Buildings Corporation
Bromley Companies
Buildings in Use
Burt Hill Kosar Rittelmann Associates
California Energy Commission
California Institute for Energy Efficiency
Carnegie Mellon University
Carrier Corporation
CEDRL, National Resources Canada
Center to Protect Workers' Rights
The Chattanooga Institute
CH2M Hill
City and County of San Francisco
City of Oakland
City of Seattle
Con Edison Solutions
Constructive Technologies Group
Cornell University
Design Harmony
Don Prowler & Associates
Durst Organization
Earth Day New York
Electric Power Research Institute
Eley Associates
Energy Center of Wisconsin
Energy Management Solutions
Engineering Resource Group
Enron Energy Services
ENSAR Group
Exergy Partners
F.W. Dodge/McGraw-Hill
FEMP U.S. Department of Energy
Fire-To-Ice
Fox & Fowle Architects
Gas Research Institute
Gensler Associates
Georgia Institute of Technology
Geothermal Heat Pump Consortium
Halton Group
Haworth
Hayden McKay Lighting Design
Herman Miller
Heschong Mahone Group
Hewlett Packard
Honeywell
IBACOS

ICF Kaiser
Illuminating Engineering Society of
    North America
Institute for Market Transformation
Interface Research Corporation
International Alliance for Interoperability
International Brotherhood of
    Electrical Workers
International Facility Management
    Association
John A. Clark Company
Johnson Controls
Lawrence Berkeley National Laboratory
Lighting Corporation of America
Marinsoft
Susan Maxman, Architects
McClure Engineering Associates
William McDonough and Partners
Montgomery County Government
National Institute of Standards and Technology
National Renewable Energy Laboratory
National Research Council Canada
Natural Resources Canada
New York State Energy Research and
    Development Agency
North American Insulation Manufacturers
    Association
Oak Ridge National Laboratory
Oakland Redevelopment Agency
Oberlin College
OmniComp/Enron
Ove Arup & Partners Consulting Engineers
Owens Corning
Pacific Contracting
Pacific Energy Center
Pacific Gas and Electric Company
Pacific Northwest National Laboratory
Passive Solar Industries Council
Portland Energy Conservation
Price Waterhouse
Prime Group Realty Trust
Public Solutions
The Real Estate Roundtable
Real Estate Technologies Group
Renewable Energy Policy Project
Rocky Mountain Institute
Rudin Management Company
Sachs and Sachs
SC Johnson Wax Company
Sequoia Architecture Group

Seventh Generation Strategies
Siemens Building Technologies
Solar Design Associates
Southern California Edison
SRC Systems
Stanford University
Steven Winter Associates
Sustainable Buildings Industries Council
Taylor Engineering
Texas A&M University
U.S. Army Cold Regions Research and
    Engineering Laboratory
U.S. Army Construction Engineering
    Research Laboratory
U.S. Department of Commerce

U.S. Department of Energy
U.S. Department of State
U.S. General Services Administration
U.S. Green Building Council
United Brotherhood of Carpenters
United Technologies Research Center
University of California, Berkeley
University of Colorado
University of Massachusetts
The Urban Land Institute
Visionwall Technologies
Walt Disney Imagineering
The Weidt Group
York International Corporation

Chapter 4

Residential Building Technology Roadmap

A Glimpse the Future

By 2020... residential building envelopes will be:

- *Energy-positive*—minimizing energy use; providing heating, cooling, and electricity; and storing or returning excess electricity to the grid.
- *Adaptable*—designed for movable walls, convertible rooms, and flexible systems to accommodate the changing needs of occupants.
- *Affordable*—cost-effective in terms of comprehensive home ownership, spanning first cost, maintenance cost, life-cycle cost, and resale value.
- *Durable*—offering enhanced safety and resistance to natural hazards, including moisture, fire, and disaster, as well as decreased maintenance.
- *Environmental*—harmless to the natural environment, resource-efficient, and appropriately balanced between embodied energy and durability.
- *Healthy and comfortable*—harmless to the well-being of construction workers and occupants and providing good air quality and flow, thermal and visual comfort, natural ventilation and light, and protection against fire, moisture, chemicals, radon, and noise pollution.
- *Intelligent*—using advanced sensors, monitors, controls, and communication technologies to improve resource efficiency, comfort, affordability, adaptability, and durability.

INTRODUCTION

This chapter represents the results of a technology roadmapping activity undertaken by representatives from many segments of the residential building envelope industry. The main question this roadmap attempts to answer is: How can the buildings and construction industry provide building envelopes that meet the demands of the consumer while reducing impacts on the natural environment?

This chapter focuses on residential buildings, including new and existing low-rise multifamily dwellings, as well as townhouses and single-family detached homes. These structures are often of similar construction type, design, and materials. Manufactured housing is not directly addressed, given its profound structural and operating differences from site building. However, many of the technical activities listed in this roadmap will yield products that can be applied in the manufactured housing market.

The main purpose of this roadmap is to define the industry's long-term vision and identify strategies to accomplish that vision. However, other results also emerge from this roadmap. First, the roadmap vision helps focus both public and private RD&D investments on industry's highest priorities. Second, the roadmapping process itself facilitates more effective partnerships between industry and government, ensuring that federal programs enhance, but do not duplicate, industry efforts. As a result, the roadmap identifies the public-private work that must be done to achieve the vision, while also building an institutional mechanism for collaboration.

WHAT THE BUILDING ENVELOPE DOES

The building shell—that part of the building that serves as an interface between the interior and exterior environments—includes the foundation (vertical wall and horizontal slab), the above-grade wall, and the roof assembly. It plays a critical role in solar gain management, thermal load control, air infiltration and exfiltration, ventilation, moisture management, fenestration support, impact and disaster resistance, noise control, air quality management, design value, and aesthetic definition.

THE ROADMAP PROCESS

The development of this technology roadmap occurred over the course of a year. During that year, over 300 industry and public sector representatives were pulled together to identify a building envelope vision for year 2020 and the strategies needed to accomplish that vision. Interactions occurred over three major workshops, numerous smaller work group meetings, and a survey instrument.

The federal government was instrumental in this process. The quality and quantity of the nation's residential housing stock impacts almost all citizens in one way or another. Facilitation of the roadmap meetings and documentation was performed by the U.S. Department of Energy (DOE).

The process for developing the roadmap was as follows:

- An **executive forum** was held in January 2000. This forum included 40 senior representatives from the building envelope industry. The participants summarized the context and history of the building envelope industry, identified trends and drivers affecting the market, and developed a draft vision of building envelopes in the year 2020.

- The executive forum was followed by a **workshop** in June 2000. At this workshop representatives from the building envelope industry refined the draft vision, identified and prioritized barriers to the vision, and defined actions for overcoming key barriers.

- From June through September 2000 a **mail and internet-based survey** was conducted among 56 senior representatives from the building envelope industry. This effort identified 120 unique technical research activities, grouped into 35 categories and ranked by risk, contribution to the vision, and potential for addressing key barriers.

- During June through October 2000, four groups of 5-10 building envelope industry representatives were formed and held **biweekly conference calls**. The outcome of these calls was a list of draft recommendations to address key barriers.

- Finally, in December 2000, 50 building envelope industry representatives reviewed and fine-tuned the draft roadmap, prioritized strategies, and developed a time frame and action plan.

STATE OF THE INDUSTRY

Trends in Perspective

Fueled by the strong economy of the 1990s, home ownership in the U.S. set a new record at 66.8% in 1999 (U.S. Census Bureau). The market for home improvements and repairs has grown 1.8% annually over the last 15 years; sales of many building envelope products have increased as much as 40% since 1990.

A number of trends are expected to impact the building envelope industry over the next 20 years. Market trends indicate a shift toward mass customization—modular homes, prefabricated building components, and do-it-yourself improvements and repair products have all experienced significant growth in market share over the last decade. The shift toward time-saving standardization and mass customization of building envelope components is likely to continue.

The Business of Building Envelopes

Despite its economic success, it has been difficult for the U.S. construction industry to develop and deploy innovative technologies and processes. A primary cause is the extreme complexity of the industry. Hundreds of thousands of companies of all sizes design, build, finance, equip, repair, and retrofit residential building envelopes. Collaboration and communication among these companies is often difficult. In addition, first cost, codes, and product familiarity have historically driven the technologies employed in the industry. This technology roadmap is an important first step in uniting the industry to overcome some of the factors that hinder innovation.

A Vision of a Sustainable Future

One of the first tasks of the roadmap participants was to develop a "vision" for the year 2020. The following vision was identified and refined in early meetings among public and private stakeholders:

Vision Statement
In 2020, building envelopes will be energy-positive, adaptable, affordable, environmental, healthy and comfortable, intelligent, and durable.

Each of the terms in the vision is important and is further explained below.

Adaptable—In the 2020 vision, the external envelope design will incorporate the adaptability or flexibility of the home, and the practical "buildability" issues associated with adaptable design will be resolved. For example, the building envelope will allow:

- Rooms that convert easily from one use to another (e.g., bedroom to office);
- Modular components that allow for movable walls;
- Features that allow the system to "grow" as the demographics of the inhabitants change (e.g., aging-in-place); and,
- System components that are easily adaptable for the future use of innovative technology.

Affordable—The vision for 2020 is one where informed consumers base their decisions upon several home ownership costs. The building envelope of 2020 should be affordable in terms of first cost, maintenance cost, life-cycle cost, and resale value.

Durable—In 2020, the building envelope will be more durable and resistant to natural hazards, offering occupants increased safety and decreased maintenance. The building envelope will better withstand moisture, fire, and disaster, and be designed with the structural strength appropriate to its geographic location.

Energy-positive—In 2020, the building envelope will minimize heating, cooling, and lighting loads through integrated design and meet remaining loads with non-polluting energy sources, returning excess electricity to the grid. This will save money and reduce emissions of greenhouse gases and other pollutants.

Environmental—In 2020, the building envelope will be resource-efficient, harmless to the outdoor environment, and appropriately balanced between embodied energy and durability.

Healthy and comfortable—The influence of the internal built environment on the comfort, productivity, and health of its occupants is an area

Figure 4-1. Zero-Energy Building

of increasing focus. In 2020, the building envelope will enhance air quality, airflow, natural ventilation, and lighting; protect against fire, moisture, chemicals, and radon; reduce noise pollution; and provide thermal and visual comfort.

Intelligent—In 2020, intelligent features will increase the affordability, adaptability, durability, energy efficiency, environmental harmony, and positive health impacts of the building envelop. Such features will enable the use of resources (light, water, and energy) only when and where they are needed.

BARRIERS

Market, Policy, and Technology Barriers
In the development of the Commercial Building Roadmap, participants identified barriers to achieving the 2020 vision. These barriers were classified as: *Market Barriers, Policy Barriers,* and *Technology Barriers.* A

voting system was used to identify the highest priority barriers for each of these categories. Table 4-1 lists the top 4 or 5 barriers for each of these categories.

Table 4-1. Major Barriers to the Building Envelope Roadmap Vision

Market Barriers
- Lack of building performance measurability
- Builder is the largest driver rather than consumer
- Fragmentation throughout the construction industry value chain
- Resistance to change
- Product perception based on market image, not performance

Policy Barriers
- Lack of nationally accepted building rating system
- Code acceptance, limitations, inconsistencies in costs
- Lack of insurance industry support and involvement
- Code enforcement; lack of inspection
- Financial incentives (e.g., tax policy, R&D incentives)

Technology Barriers
- Systems integration of building components and how they function
- Lack of skilled labor—acceptance of substandard work
- Lack of collaborative R&D for systems
- No process for discovery of interactive effects of new products

Overarching Barriers

In addition to the barriers identified above, several Overarching Barriers were identified. These barriers are those obstacles that hinder the entire industry from moving forward.

Barrier 1: Lack of Education/Awareness. Industry fragmentation perpetuates information fragmentation and vice versa. Researchers are not informed about industry structure and marketplace needs. Product manufacturers, architects, designers, engineers, builders, contractors and code officials need to be able to technically assess the characteristics of new products, materials, and systems. They also need to know proper installation and operating methods. Consumers and homeowners need information that will enable them to generate demand for the 2020 vision and influence builders' decisions.

Barrier 2: Non-systems approach to building envelope construction. Many failures in building envelopes occur at the interfaces of products made by different companies or manufacturers and installed or repaired on-site by a variety of different trades. While the individual parts in complex building envelop assemblies must retain their integrity, interface complexity needs to be simplified. Development of clear assembly protocols for installation, replacement, and repair of subsystems would benefit all industry segments. A wealth of innovative materials, systems, and processes is currently available for enhanced building envelopes, but these resources are not utilized as fully as they could be. Collaborative R&D would establish a solid platform on which a systems approach to the design and construction of the building envelope could be established. Collaborative R&D should be emphasized within each vision element, with cognizance of the potential synergies and pitfalls created by the interactions between the vision elements.

Barrier 3: Shortage of skilled labor. The shortage of skilled labor within the construction industry has been exacerbated by the combination of high construction demand and low national unemployment. Builders are increasingly forced to employ inexperienced workers, which lengthens the construction schedule and may lead to improper installation of components. Even though the number of skilled workers has grown 5% annually since 1990 (3.9 million in 1999) an additional 240,000 skilled workers per year are needed (Knight Ridder/Tribune Business News, February 29, 2000).

Barrier 4: Absence of total system performance measurement. Consumers are faced with an overwhelming amount of information, multiple experts, and a short decision-making time frame when purchasing a home. No comprehensive guide exists to support the choices of consumers making this significant decision. A simple, user-friendly, voluntary rating system is needed to assess and communicate the performance of a building envelope on multiple attributes (e.g., the elements of the 2020 vision). The rating criteria should be objective and scientifically based.

Such a metric analysis product would enable consumers and designers to make more intelligent decisions and would serve as a powerful marketing tool for builders and building product manufacturers.

Barrier 5: Difficulty for new and emerging technologies to achieve building code acceptance. To protect public health and safety, building regulations establish minimum criteria for building designs and construction. Regulations are based on model building codes developed in the voluntary sector, principally by three regional organizations: Building Officials and Code Administrators International, Inc. (BOCA); the International Conference of Building Officials (ICBO); and the Southern Building Code Congress International, Inc. (SBCCI). The International Code Council (ICC), made up of the previous three, will eventually become the single nationwide building code, replacing the three regional model codes. Currently, the issue is adoption and enforcement of such a code by state and local governments on a uniform basis, which will occur over the next few years as ICC codes are adopted in place of regional codes. Model codes tend to be prescriptive, but they accept new designs, processes, and products if code requirements are met on the basis of equivalent performance. However, the lack of product testing, standards, and familiarity can inhibit the approval process.

STRATEGIES

Achieving the vision will require a parallel strategic approach: *market/policy strategies* to surmount key barriers to innovation and *technology strategies* to exploit expanded use and new direction in materials, products, systems, design processes, and construction practices. The five market/policy strategies presented here were developed by industry working groups, while the detail for the one overarching technology strategy was gathered from a broader industry group through an industry survey. Each of these strategies is presented below.

Strategy 1: Promote Education/Outreach along the Construction Value Chain

This strategy attempts to overcome the barrier presented by a lack of education and awareness in the industry, which causes industry fragmentation. The strategy should pursue the goal that all members of the construction supply chain make informed decisions aligned with the 2020 vision.

This strategy includes specific suggestions, such as:

- Develop and distribute to industry up-to-date envelope design information, packaged in best practice guides and knowledge tools to transfer ideas;
- Develop a building envelope rating system for marketing to consumers;
- Increase dialogue between industry members and code officials to ensure acceptance of technologies by appropriate codes;
- Increase collaborative R&D as a platform for systems approach;
- Develop a "Teach the Teachers" program for designers, architects, engineers, builders, and contractors;
- Support development of continuing education programs for professionals in energy technologies and building intelligence;
- Develop and disseminate a procurement strategy that rewards architecture and engineering for the additional analysis and design effort needed to achieve superior building performance;
- Publish "plain talk" guides to best practices for envelop construction;
- Educate homeowners on how to prolong the life of home building materials;
- Educate homeowners about energy consumption;
- Use television programs to link homeowners with current research.

Strategy 2: Build a Platform for Collaboration, Leading to a Systems Approach and Improved Envelope Construction

This strategy is aimed at overcoming the "non-systems" approach that currently exists in the construction industry. The desired outcome of this strategy is to increase collaborative R&D leading to system approaches and improved envelope construction. Mid- and long-term strategies that should be pursued are shown in Table 4-2.

Some specific areas identified for collaborative R&D efforts include:

- Intelligent building materials (advanced polyvalent materials, self-adjusting/repairing materials);
- Design for Adaptability (product design for assembly/disassembly, replacement/repair, recyclability)
- Disaster resistance;

- Envelop component integration (product interfaces, component integration);
- Moisture (detection and resistance);
- Design tools (CAD);
- Resource-efficient materials.

Table 4-2. Mid- and Long-term Strategies for Increasing Collaborative R&D

Improve environment for collaboration:
- Create environment "friendly to collaborative R&D" based on best practices from case studies, barriers, and incentives.

Conduct R&D:
- Prioritize and begin R&D.
- R&D should be coordinated nationally and have a regional focus.

Technology transfer:
- Develop knowledge products that synthesize output of collaborative R&D and can be used by a mass audience (e.g. best practice guide).
- Link knowledge products to training.
- Technology transfer should be coordinated nationally and have a regional focus.

Strategy 3: Expand the Skilled Workforce Trained in Labor-reducing Technology

This strategy is aimed at overcoming the shortage of skilled labor. The outcome from this strategy is building an adequately skilled workforce trained in labor-reducing technology. Mid- and long-term strategies that should be pursued regarding developing a skilled workforce are shown in Table 4-3.

Expanding this skilled labor pool is a broad issue involving needs and interests from many quarters. Numerous public and private organizations are dedicated to developing the U.S. workforce. One large opportunity to increase skilled labor involves high-profile career recruitment, in collaboration and coordination with existing skilled labor organizations (such as, Associated Builders and Contractors, Consortium of Community Colleges, Home Builders Institute, the U.S. Department of Education, and the National Center for Construction Education Re-

**Table 4-3. Mid- and Long-term Strategies for
Developing a Skilled Workforce**

Increase attractiveness and entry into trades:
- Work with all levels of education: high school, apprentice pro-
  grams, vocational schools, existing labor, supervisors.
- Build image of trade as career choice.
- Determine best practices and strategy for labor recruitment.

Develop labor-reducing technology:
- Conduct research and development of technologies for retrofit and
  new construction markets.

Develop training and third-party certification:
- Working with specific trade association and key stakeholders
- Develop curriculum and "hands-on" skill development in accor-
  dance with best practices.
- Develop "instructor version" of curriculum for "train the trainers"
  program.
- Determine third party to perform certification and define criteria
  for certification.
- Provide incentives for participation in training and certification.

---

search, among others).

Along with recruitment, a concentrated effort to deploy technolo-
gies that reduce and/or simplify on-site labor could be undertaken, lead-
ing to enhancement of the image of the trades as well as better quality
assembly and build processes. Accelerating RD&D on the technologies
below could lead to better labor capabilities at building sites:
- Advanced framing
- Modular coordination
- Envelop component integration
- Diagnostic tools
- Design tools

Support and incentives—in the form of grants, tax breaks, insurance reductions—could also be offered to builders and contractor to
encourage the use of trained labor or to adopt a training program.
Higher wages and better benefits will also be key to attracting and keeping career-minded skilled trades workers.

Strategy 4: Develop a Building Envelope Performance Rating System
This strategy is focused on overcoming the absence of total system performance metrics. The expected result from successfully implementing this strategy is a widely accepted set of tools that provides easily understood information on the performance of the building envelope and allows the performance rating of the building. Mid- and long-term strategies that should be pursued regarding developing a building envelope performance rating system are shown in Table 4-4.

This roadmap identifies a number of potential performance rating system attributes, found in Table 4-5. These attributes could be integrated into a Consumer Report style that is easy to use, inexpensive, and acceptable to industry, appraisers, and insurers.

Strategy 5: Support the Acceptance of
Emerging Technologies by Codes and Standards
This strategy attempts to overcome the difficulties faced by new and emerging technologies to achieve building code acceptance. The results from this strategy would lead to quicker adoption and acceptance of emerging technologies. Mid- and long-term strategies that should be pursued regarding supporting acceptance of emerging technologies by Codes and Standards officials are shown in Table 4-6.

A number of efforts can be made to support the acceptance of emerging technologies by Codes and Standards officials. Many of the technical areas identified for research and development will not win immediate regulatory approval. First they must be compared to similar (already approved) technologies in order to determine which codes and standards will be applicable. Existing resources should be mined for methodologies that could support building code acceptance of new technologies. For example, a nationwide building code incorporating both prescriptive and performance-based standards should be implemented, assuming testing protocols could be developed to help evaluate new technologies using performance based mechanisms.

Strategy 6: Develop, Evaluate, and Promote the Adoption of Building Envelope Materials, Systems, and Process/design Techniques, Aligned with One or More of the Vision Elements
This general strategy is designed to pull the entire construction and design industry together to help achieve the 2020 vision from a technological standpoint. In the near-term (3 years) the strategy calls for:

**Table 4-4. Mid- and Long-term Strategies for
Developing a Performance Rating System**

Continue building audience:
- Engage key stakeholders throughout process to ensure acceptance.

Develop rating system:
- Determine elements of existing systems that satisfy desired attributes and capabilities.
- Define and begin research needed to fill gaps between existing and desired rating system.
- Synthesize desired features into comprehensive, technically accurate rating system.
- Rating system should be developed one element at a time (with priority elements first).
- Elements should have a consistent "output" or metric.

Technology transfer:
- Develop a transparent, accessible, simple-to-use set of tools and information
- Provide incentives or requirements to participate.
- Provide a clear distinction and relationship to other rating systems.
- Begin a well-funded, coordinated market transformation program.

Collect data:
- Collect appropriate data for system attributes from manufacturers and industry groups.
- Assemble data into database of products, assemblies, and construction type (via graphical interface).
- Fill gaps in data through calculation and testing.
- Make database available to users.

***

- Initiating joint industry/government technology implementation steering group;
- Determining high-priority technical activities;
- Investigating current market status of priority technology activities;
- Increasing research funding for priority technical activities;
- Developing a critical 20-year technology and implementation plan

Table 4-5. Attributes and Metrics for a Performance Rating System

Vision Attribute	Suggested Measurement
Adaptable	• Ability to change components/rooms • Designed for adaptation
Affordable	• Monthly housing cost • First cost • Maintenance cost • Life-cycle cost
Durable	• Performance of components and systems by climate • Resistance to wind, earthquakes, fire, weathering, pests, maintenance
Energy	• Energy efficiency: energy consumption per month
Environment	• Life-cycle assessment for embodied energy and raw materials
Health/Comfort	• Indoor air quality: mold, radon, lead, volatile organic compounds, toxics • Acoustics • Thermal comfort
Intelligent	• Degree of interaction (sensor/reaction) between house components, occupants, and external environment

***

pathway to produce a balanced portfolio of priority technical products.

In the longer term (10-20 years) the strategy calls for:
• Conducting technology transfer activities for all materials, systems, tools, and performance evaluation results;

**Table 4-6. Mid- and Long-term Strategies for
Supporting Codes and Standards Acceptance**

Develop test procedures:
- Provide a resource for gaining appropriate tests.

Develop standards:
- Provide a resource for development of appropriate standards.

Ensure code acceptance:
- Work with product designers/manufacturers and code officials throughout product development to educate and ensure acceptance of technologies by appropriate codes.

***

- Encouraging industry partners to manufacture and market materials and systems once a technology has reached the pilot/demonstration stage;
- Developing, testing, and promoting the adoption of materials and system aligned with one or more of the vision elements through RD&D;
- Developing information outreach thrust for the deployment of process and design tools;
- Evaluating performance of existing and new emergent materials and assemblies.

The important technology research identified by stakeholders involved in this roadmap, fall into four RD&D areas (a detailed list is included in the Appendix to this chapter):
- *Materials*—potential for greater energy efficiency, control performance, cost reduction, environmental consideration, and ease of construction and maintenance through innovative materials.
- *Systems*—Addresses innovations within envelop components and assemblies, considering the envelope as a complex, multi-layered, integrated part of the building.
- **Design and construction process**—Technologies and processes that minimize materials, energy, time and expense through better design, specification, construction, and operation and maintenance of the building envelope.

• *Performance evaluation*—Processes to successfully conduct and manage modeling, monitoring, and rating the performance of the building envelop with regard to moisture, solar radiation, temperature differential, pressure differential, wind pressure, and impact and catastrophic failure.

NEXT STEPS

This chapter outlines an ambitious vision for the residential buildings envelope industry. It serves as a resource for the public and private sectors and offers a framework for greater collaboration across the industry in creating new market opportunities for high-performance commercial building. The roadmap is also useful in providing guidance for future private and public research and development focus.

Over the next few years, industry will work to implement each strategy. The implementation process will follow these general guidelines:

• Identify leadership, recruit champions, establish steering group for each strategy

Figure 4-2. Construction Particle Board Made from Wheat Straw

- Develop strategic vision, mission, goals and target audiences
- Research existing best practices and resources
- Identify gaps to be filled
- Ensure continuous improvement through monitoring and feedback cycles and adjustments.

The roadmap intentionally excludes detailed implementation approaches. Such approaches will be jointly developed over time between government and industry. For readers interested in following the process of this particular roadmap, please go to: *http://www.eere.energy.gov/buildings/research/roadmaps.cfm*.

[**Editor's Note**: This chapter represents an abridged and edited version of the U.S. Department of Energy report entitled *Building Envelope Technology Roadmap* published by the Office of Building Technologies, U.S. DOE, May 2001. The editor is grateful to the U.S. Department of Energy for permission to publish an abridged version of this report.]

APPENDIX 4-1:
ACKNOWLEDGEMENTS FROM THE RESIDENTIAL BUILDING TECHNOLOGY ROADMAP DOCUMENT

A NOTE FROM ...
   Ronald Santoro, Team Leader, *Building Envelope Technology Roadmap 2020*.
   As team leader of the roadmap development process for DOE's Office of Building Technology, State and Community Programs (BTS), I wish to thank the hundreds of organizations and individuals that contributed to this significant effort:

AFL-CIO Building and Construction Trades Department
American Architectural Manufacturers Association (AAMA)
American Forest & Paper Association (AFPA)
American Portland Cement Alliance (APCA)
American Solar
Andersen Corporation
Aspen Research

Ball State University
BP Solarex
Brick Industry Association
British Columbia Building Envelope Research Consortium (BC-BERC)
Building Technology
Building Technology Group, Department of Architecture, MIT
Cellulose Insulation Manufacturers Association (CIMA)
Celotex

Center to Protect Workers Rights (AFL-CIO)
CertainTeed Corporation
Champion Enterprises, Inc.
Commonwealth of Massachusetts Division of Energy Resources
Corbond Corporation
Construction Technology Laboratories, Inc. (CTL)
DAP, Inc.
Department of Business, Economic Development, and Tourism (DBEDT), State of Hawaii
Development Center for Appropriate Technology
David L. Roodvoets Consulting (DLR) (representing the National Roofing Contractors Association)
DMO Associates
The Dow Chemical Company
DuPont Tyvek Weatherization Systems
Energy Services Group
The Engineered Wood Association (APA)
Engineering Field Activity Chesapeake, Naval Facilities (EFACHES, NAVFAC)
Exterior Insulation Manufacturers Association (EIMA)
Florida Solar Energy Center (FSEC)
General Electric Company
Grace Construction Products
Habitat for Humanity
Icynene, Inc.
Innovative Design
Institute for Research in Construction (IRC)
Intech Consulting, Inc.
Jeld-Wen, Inc.
Johns Manville
Louisiana Pacific
Manufactured Housing Institute
Marvin Windows
Masco Corp.
National Association of Home Builders (NAHB)
National Association of Home Builders Research Center (NAHBRC)
National Concrete and Masonry Association (NCMA)
National Institute of Building Sciences, Building Environment Thermal Energy Council (NIBS, BETEC)
National Institute of Standards and Technology (NIST)
New York State Energy Research and Development Authority (NYSERDA)
North American Insulation Manufacturers Association (NAIMA)
Ohio Office of Energy Efficiency
Owens Corning
Partnership for Advancing Technology in Housing (PATH)
Pennsylvania Housing Research Center
Portland Cement Association (PCA)
Rock Wool Manufacturing
Rutgers Cooperative Extension
Seattle Department of Design Construction and Land Use
South Carolina Institute for Energy Studies
Spray Polyurethane Foam Alliance—America Plastics Council (SPFA-APC)
State Service Organization (SSO)
Superior Walls of America, LTD
Sustainable Building Industry Council (SBIC)
Texas A&M
Trex Company
U.S. Department of Energy (DOE)
U.S. Department of Housing and Urban Development (HUD)
United States Gypsum Corporation (USG)
University of Minnesota, College of Architecture & Landscape
University of Waterloo
Virginia Tech
Washington State University
Weatherization Assistance Program—Technical Assistance Center
What's Working (representing the U.S. Green Building Council)
Window and Door Manufacturers Association (WDMA)

Special thanks to Ed Barbour and Kathryn Fry, of Arthur D. Little, for their role in organizing and documenting the roadmap process and creating the first roadmap drafts;

Doug Brookman, of Public Solutions, for his creative facilitation of the three workshops; and the team at Brandegee for their vision and creativity in bringing the roadmap together into the document you see today.

APPENDIX 4-2: DETAILED TECHNICAL R&D ACTIVITIES

The items below represent a short description of important research and development activities identified by participants in the residential building roadmap workshops. These R&D ideas will help guide public and private sector R&D activities over the next 20 years.

MATERIALS
Advanced Aggregate Materials
- Develop innovative aggregate substitutes for weight reduction and thermal insulation in cast-in-place and precast concrete.

Advanced Insulation
- Develop environmentally friendly super-insulations.
- Develop high-thermal/low-volume, durable, energy-efficient envelope components.
- Develop simple installation systems for advanced (high R-value) insulation.
- Develop economical manufacturing processes for advanced (high R-value) insulation systems.
- Sponsor research to provide a better understanding of R-values in aged insulation materials.

Air/Vapor Barriers
- Develop air barrier concept integrated into thermal envelope assembly.
- Develop a diodic vapor barrier for use in mixed climate regions.
- Develop a spray-on coating for building joints/seams that restricts water/air penetration and is easy to apply.
- Develop an easy-to-use factory-applied gasket to air-seal multisection HUD code or modular homes. Gasket to seal 1/4" to 1/2" gaps.
- Develop OSB (oriented strand board) which will have high vapor permeability and high decay resistance.

Cellular Building Components
- Research latent performance attributes of cellular morphologies in a variety of building components, including exterior wall assemblies, interior wall surfaces and substrates, roof assemblies, floor assemblies, structural components, and other elements.

- Explore the benefits and performance attributes of various raw materials that can be used to fabricate advanced cellular building components. Evaluate the potential of cellular morphologies in the areas of aggregate materials, insulation, wood, air, and vapor barriers, and disaster-resistant materials.

Disaster-Resistant Materials
- Develop ignition-resistant super-insulation materials for building envelope.
- Develop an interior finish wall panel of a composite material class with embedded fibrous or other fire-monitoring material capable of detecting fire spread and alerting residents. Coordinate interior wall panel and central fire alarm or distributed alarm systems. In this instance the central alarm system would be made redundant. All fire detection and alarm functionality would be satisfied by the panel itself.

Fabric Technology
- Develop fabrics composed of capillary fibers of self-healing resins designed to form patches over areas that have been damaged.
- Invest in research toward the use of fabrics within orthogonal, panelized frameworks with integral insulation, and explore the possibility for on-site rigidizing fabric surface.
- Study impact resistance of high-strength fabrics and various "woven" technologies for use as exterior sheathing or interior reinforcement.
- Counteract positive and negative air pressure with fabric wall reinforcement.
- Reduce or eliminate catastrophic failures with selective textile reinforcements of wall and roof assemblies.

Intelligent Materials (Self-Adjusting/Regulating/Repairing/Polyvalent)
- Develop self-righting structures and other types of "responsive" structures to eliminate damage to exterior wall and interior finishes due to differential settlement and other types of short- and long-term building movement.
- Create a material and/or assembly with variable gas permeability that adjusts to exterior environmental conditions. User control bypass is also an important performance requirement.

- Produce variable R-value insulating material capable of adjusting the interior insulation's effective R-value, thereby allowing heat transfer to be regulated by changing diurnal peaks.
- Develop self-healing building materials—materials that have a variety of self-healing properties, that is, the ability to repair damage and or inconsistencies that could compromise the performance of that material. Self-healing materials could minimize construction and repairs waste, improve quality through "embedded" control mechanisms and facilitate crews in the tasks of detecting failures or problems.
- Create self-healing insulation materials—evaluate the combined use of expanding agents with cellular materials to provide self-healing properties for the insulation matrix.
- Design wall assemblies and details that effectively collect incidental moisture into a safe storage location or dispose of it safely.
- Develop highly polyvalent composite-material systems able to satisfy complex and multivariate performance requirements of architectural assemblies.
- Create a series of modular wall units displaying variable thermal attributes while maintaining or increasing their durability.
- Develop classes of polyvalence materials with embedded information-technologies functionality.

Moisture-Control Materials
- Develop building materials and/or products that can be applied to inhibit mold growth.
- Develop materials to provide cladding and protection under different climatic conditions.
- Design wall systems with improved resistance to water penetration.
- Develop breather-type sheathing membranes to act as moisture and air barriers, have a 100-year service life, and effectively integrate with other components of the building envelope.

Nontoxic Materials
- Create new materials that reduce or eliminate allergic and other reactions among people with sensitivities and thus improve indoor air quality.

Radiant Technologies
- Develop, refine, and promote radiant barrier technology for residential wall assemblies for hot climates.
- Develop better radiant metal deck assemblies.
- Test various types of radiant barriers in roofs and attics in combination with various shades of roof color. Document effectiveness and payback times.
- Research impacts of radiant heat loads in attics.

Resource-Efficient Materials
- Produce panelized enclosure systems composed of alternative and/ or recycled materials such as recycled plastics, straw, and fly ash combined with traditional materials such as concrete, wood, or steel.
- Promote the development and adoption of panelized building materials that combine structural and energy-efficient properties with the use of recycled or waste resources such as waste agricultural products; e.g., Strawmit brand compressed straw panels. Research solutions to issues such as moisture- and fire-resistance.
- Develop recycled roof coverings with performance attributes comparable to or higher than those of new products or materials.
- Produce alternative materials using recycled and/or waste material such as:
  - Viable exterior wall panels composed of cellulose and lignin-based natural fibers reinforcing a natural resin matrix (e.g. agriculture fiber composite exterior wall panel).
  - Lightweight natural-fiber-reinforced cementitious panels for stress-skin exterior wall applications. Natural fibers—such as straw, wheat, and other agricultural products—provide structural and thermal resistance properties.
  - Biomaterial-mediated exterior wall and roof assemblies. Support research focused on the potential integration of natural or "green" materials into exterior wall and roof assemblies of small structures, particularly residential buildings.
- Develop guidelines and standards for the performance requirements of recycled materials and products.
- Research alternative processing methods and practices that can reduce the cost of recycled materials.

SYSTEMS

Advanced Foundations
• Develop foundations that are low energy losers, impervious to moisture and insects, natural disasters, and a variety of soil conditions.

Advanced Panel/Prefabrication
• Develop building enclosures and structural systems composed entirely of 2- and 3-dimensional composite panel and structural element technology.
• Support for projects that eliminate the proliferation of discrete building components and minimize complexity while maximizing the variation and flexibility, durability, and cost-effectiveness of building systems utilizing composite panel technology.
• Create automated manufacturing techniques to mass-produce wall panels of new or nontraditional raw materials.
• Develop pre-manufactured building enclosure panels. Panels might be designed with an underlying modular dimensioning approach to minimize waste of materials while allowing for easy handling by two workers.
• Research the downstream costs of call-backs, warranties, repair, and upkeep resulting from excessive dependence on on-site labor for installation of separate parts, in contrast to prefabrication at off-site plants.

Crawl Spaces
• Research effective insulation techniques for open-web truss floor systems.
• Research energy, moisture, and IAQ impacts of vented versus unvented crawl spaces.

Double Envelope
• Develop glass and opaque double-envelope systems with integral energy collection and distribution. These systems combine two or more layers separated by one or more cavities for the purpose of collecting and utilizing (winter mode) or rejecting (summer mode).

Energy Services/Supply
• Develop building dehumidification using heat from building envelope.

- Develop roof and/or wall assemblies that collect and supply energy to the building envelope.
- Develop wall systems that integrate photovoltaics, similar to new roof shingles.
- Develop roofing systems that produce energy through PV-capture heat for space heating or water heating.
- Integrate photovoltaic circuiting, piping, waterproofing, insulation, and textured interior surface.
- Develop thermally driven air conditioning system using solar heat from the envelope.
- Research technologies for the production of energy-generating exterior walls.
- Develop simple, mass-producible wall and roof systems that can act as solar thermal collectors.
- Develop nontraditional solar thermal space heating using the building envelope.
- Generate low-cost thermal storage with air heat exchanger using the building envelope.

Envelope Component Integration
- Create enclosure assembly techniques that go beyond conventional systems interfacing to true integration. For example, wall assemblies that also serve as HVAC system supply or return ducts; foundations that can be part of an on-site rainwater collection and storage system.
- Develop window, door, and other envelope-penetration systems that are leak-proof and reduce envelope's discontinuities.
- Develop improved industry standards for the full range of building envelope product interfaces through collaborative R&D and targeted economic incentives.
- Develop effective, environmentally acceptable joint closures (e.g. window/wall interfaces; durable, air-tight sealing mechanism for exterior door).
- Create innovative building envelope (walls, windows, and roof) components to optimize the energy-efficiency of buildings in different climate zones.
- Support projects that simplify construction of the envelope and its components.

Intelligent Envelope Systems
- Develop intelligent object data format standards, in partnership with NIST. Goal is standard for data type, location, to facilitate seamless integration between DOE 2/blast and CAD data.
- Research future trends of Smart Home systems and predict customer needs in the next 10 years and their impact on the building envelope.
- Develop sensor and control (embedded and otherwise) technologies and techniques for intelligent enclosure systems.
- Create active surface (inside) temperature control: MRT control which allows reduction of T-stat setting in winter, resulting in potentially lower energy use winter and summer while increasing comfort.

Rain Screen
- Develop, refine, and promote rain screen technology for residential wall assemblies.

Roof/Attic Systems
- Promote R&D to minimize arbitrary decisions in roof maintenance and repairs. The following areas are suggested:
  — When to leave a roof in place and "reuse" the insulation
  — Studies and dissemination on roof drying
  — Tools that tell when to re-rof and when to re-cover shingles
  — Life performance of recycled versus new roofing materials
  — Increased use of design tools for wet/problem areas
- Research capability problems with ice dams and moisture.
- Research potential benefits of roof insulation and conditioned attics versus ceiling insulation.
- Develop advanced roofing materials that can increase heat rejection (avoid heat islands) and environmental efficiency (run-off from asphalt shingles; green buildings).
- Increase workers' protection through the development of new materials and/or new systems; e.g. systems that reduce worker exposure to particulate and gaseous materials; materials that are less toxic/irritating.
- Intensify research of living roofs and their potential benefits.

Super Walls
- Create various kinetic wall assemblies to increase the integration

between interior and exterior spaces.
- Develop wall systems that are low energy losers but, more important, are impervious to moisture, water, air infiltration, and insects, and can withstand natural disasters. The design should be adaptable to different climates.

PROCESS / DESIGN

Advanced Framing
- Promote engineered framing technology (reduces amount of materials) in residential construction.
- Develop and demonstrate durable and thermally efficient building materials and more efficient framing techniques.
- Design low-cost connections between steel-framed interior walls, roof trusses, and floor joists.
- Research opportunities to modify existing wood structural building envelope materials to increase their thermal barrier properties.

Automation
- Develop control systems and robotic mechanisms for the viable implementation of on-site automated material handling, sorting, assembly, and finishing of entire building systems. Possible scenarios would be the development of various mechanical and computational systems for the on-site automated assembly of exterior walls, foundation systems, roof assemblies, and other fully functional building assemblies.

Daylight and Passive Solar Design
- Develop glare reduction devices that optimize daylighting savings potential. Glare-reduction devices located outside glazing line would reduce the need for occupants to close blinds and turn on lights, and would maintain desirable solar gain during heating seasons. The characteristics of such devices could be linked to exterior conditions through a control system.
- Improve daylighting control systems (e.g. louver systems that replace interior lighting).
- Promote the design of passive solar heating and cooling load avoidance, including building orientation and shading.

- Promote the design of buildings with roof orientations suitable for solar water heating and photovoltaics, and install plumbing chases for future solar system installation.
- Develop daylighting system that has almost no heat gain and distributes light evenly. Measure effect on occupants.
- Conduct productivity research associated with daylighting.

Design for Adaptability
- Develop design concepts for building envelopes (separate for low-, medium-, and high-rise) that will facilitate future changes from one building type to another.
- Develop new "open architecture" knowledge of product integration principles that can reduce dependencies for ease of assembly, disassembly, replacement, and repair.
- Develop "plug-n-play" components and systems that can be easily added and removed.
- Design foundations that allow for movable walls.
- Design basements that are "potentially furnishable."

Design for Intelligence
- Develop a design process for intelligent materials and systems.

Design Tools
- Develop tools that allow design professionals to perform integral analysis and design of buildings. These tools should be coupled with a drafting package for additional convenience.
- Develop CAD-based expert systems combining graphical and script-based capabilities to describe typical interfaces in building envelopes to:
  — Give up-to-date information on available products in a product database
  — Provide direct feedback about the problems that need to be solved in product interfaces
  — Give feedback on assembly performance with a given product specification
  — Support visualization of the composition and joints between products
  — Have parametric behavior of elements
- Analyze the economic and technical advantages of higher-value-

added "for-stock" components in terms of life-cycle costs of building envelopes made from such products versus conventional products.

- Continue to develop a database of life-cycle costs for residential construction.

Modular Coordination
- Develop underlying modular dimensioning strategies for enclosure systems that minimize waste during both manufacturing and construction and represent a high degree of dimensional compatibility with other building components.

Natural Ventilation/IAQ
- Develop design methods and/or materials that allow for natural, nonmechanical air movement through the building envelope.

Recycling/Reuse Processes
- Create market incentives/information support structure for the recycling/reuse infrastructure of construction materials, equivalent to U.S. cardboard recycling or German wood pallet recycling. Incentives could include policy, government procurement, and/or market development
- Develop fabrication and glazing technologies for insulated glass units, allowing renewal by disassembly and resealing when edge seals fail, so that durable materials remain in service.

Regional Design
- Research proper processes that customize the design of building envelopes to allow for regional climatic differences.

PERFORMANCE EVALUATION

Modeling/Testing
Energy
- Create a database and design tool for roof designers that tie all "energy effects" of roof products and assembled roof. Some performance testing will be needed to fill data gaps (how color, mass, and insulation affect each other).

- Develop a building modeling program that adequately predicts energy losses through roofs so roof systems can be compared, as well as through basements and foundations so those systems can be compared.

Monitoring/Testing
- Use existing organizations that have testing protocols for developing testing systems, new protocols, and national standards. Industry leaders need to be involved in pushing for industry standard testing in industry group meetings and as a competitive advantage in the marketplace. This will drive all competitors to standards.
- Develop diagnostic tools to measure and evaluate proper installation of various envelope components.

Disaster Resistance
- Create wind and hail standards and design guides for roofs. In addition, data for a variety of roof types (steep and flat) at various speeds and pressures is needed.
- Develop a test missile that simulates hail and can be effectively used on both soft and brittle materials.
- Design a test that evaluates all discontinuous products for steep roofs and one that evaluates dynamic performance of flat roofs.
- Determine response of panelized walls to earthquake forces.
- Create detection and monitoring technologies to evaluate critical loads due to tornadoes, earthquakes, and other catastrophic events.

Durability
- Develop durability predictor curves based on a material's chemical, galvanic interaction with temperature, humidity, and structural performance (e.g., brick durability).
- Develop tests on the longevity 23 of roofing and other building materials. These tests should include the effects of:
  — Stress from cyclical temperature changes
  — Reflectivity
  — Temperature of ventilated/unventilated attics and decks
  — Ultraviolet, rain, and wind exposure
  — New inorganic materials
- Develop criteria for siding materials to define the level of durability of these products.

• Increase knowledge of stucco properties, determine how properties affect applications, and integrate findings into a comprehensive stucco best practices guide.

Moisture
• Formulate and validate testing of compliant roof/wall assemblies to detect moisture migration and water penetration under static loading.
• Create tools to predict condensation of moisture within building envelopes and standard details to reduce condensation in building envelope.
• Develop industrial protocol and equipment for monitoring moisture in building envelopes during and after construction.
• Develop material data for use in moisture analysis.

Rating Criteria
• Develop research criteria to improve acceptance rate for new building products and materials, so innovations are evaluated based on the merits of the new technology and not only on comparison with existing technologies.
• Develop equitable procedures to measure in-place performance of building materials. These procedures would take into account R-value, air presence, durability, energy use, health, etc.
• Combine all existing energy-rating agencies into a single body that measures building performance for a period of several years after construction.
• Develop means (such as an "energy tag") to communicate the life-cycle or annual energy cost of a building.
• Create a catalog and database containing information on the IAQ performance of sealants, adhesives, and finishes.

Chapter 5

Windows Industry Technology Roadmap

A Glimpse of the Future

By 2020... Windows will be "smart" and will operate with the surrounding building environment as a system. These windows will be extremely energy efficient, employing gels and other new materials to improve insulation. But they will also be able sense energy loads from building exteriors and interiors and adjust their insulation properties based on energy needs. Using holographic techniques, windows will be able to direct outside light to particular areas of the interior space, thus replacing artificial light with natural light in the work environment.

Windows will have photovoltaic technology embedded in them, thus not only keeping energy balanced within a building's system, but also generating power from solar energy and feeding this power to the building load.

Windows will also be entertaining and informative and will use imaging to provide information (graphical or text) displays to viewers—these displays may even be dynamic, perhaps indicating storefront sales, outside weather conditions, or local news.

Windows will be durable and shock proof and will be able to withstand significant environmental changes, but they will also be adaptable and modular and could be substituted with new advanced products easily and cost-effectively.

Finally, windows will be produced in an environmentally responsible manner, designed for recyclability, modularity, and upgradeability.

INTRODUCTION

In addition to over 400 window fabricators, the window industry includes glass manufacturers, vinyl and aluminum extruders, wood suppliers, distributors, retailers, and contractors. Serving primarily residential and commercial markets, window sales exceeded 1,200 million square feet and $7 billion in 1997. Owing to their increasing versatility, windows make up a striking portion of wall area in new construction—13 percent in new residences to 50 percent in large office buildings. The window industry of today is a vibrant, modern set of businesses and is well positioned for the challenges it will face in the next two decades.

In spite of the success of the window industry, significant challenges lie ahead. The industry must continue to meet society's changing expectations while remaining economically viable and globally competitive. Meanwhile, the window industry is in the midst of rapid technological change. Recent developments in glazing, framing, and assembly have dramatically improved the energy conservation potential and quality of new windows. This pace of technological development should continue in response to trends in new construction and retrofit that place a premium on energy conservation, enhanced quality, fast delivery, and low installed cost. Trends in the window industry, economy, and society will drive the window industry of the future, as will uncertainties and the rapid advance of technology.

This windows roadmap presents the thinking of numerous industry and government stakeholders about the future of the windows industry. Of particular focus are the energy conservation properties of windows and their ability to provide comfort and appropriate services efficiently and cost-effectively. The first part of this chapter discusses the process of this roadmap development and an appropriate vision statement developed by the windows industry. This is followed by identification of barriers to that vision, and finally strategies that can be employed to overcome those barriers.

THE WINDOWS INDUSTRY ROADMAP: VISION STATEMENT AND ROADMAP PROCESS

In September 1998, the window industry began the process of developing a technology roadmap with a one-day Executive Visioning

Forum held in Chicago. During this forum, two-dozen industry participants discussed their current situation and outlined a long-range vision
for maintaining and building their competitive market position.

Setting the context for the vision statement development was a
discussion of trends and other factors that seem to be driving the windows industry. These influences are summarized in Table 5-1.

Table 5-1. The Context for the Windows Industry Vision

Industry Trends	Industrial trends reflect vigorous competition in the construction products market: • Shift toward low-e glazing and new framing materials • Reduction in production cycle time • Increased automation • Development of systems approach to building design • Declining window prices • Consolidation of fabricators and contractors
Economic Climate	Economic trends reveal opportunities for windows to provide additional value to consumers: • Rise in disposable income • Growth in replacement market through reno-vation and upgrading • A strong economy • Low energy prices
Social Trends	Social trends hint at changes in consumer percep-tions and values: • Aging population • Increase in home ownership • Heightened environmental awareness • Increased role of women in purchases of home building products
Potential Barriers	Some important trends remain uncertain, and the industry's vision must be flexible and responsive to:

(Continued)

Table 5-1 (*Cont'd*)

Potential Barriers	• Deregulation of utilities • Housing and construction trends • Enforcement and compliance of building codes
Technology Factors	The window industry is in the midst of the greatest technology change in its 300-year history—a phenomenon that influences industry dynamics into the foreseeable future: • Opportunity for market differentiation • Accelerating rate of technology change
Regulatory Trends	Government programs and regulatory efforts have the potential to either enhance or stifle innovation: • Growing industry participation in the regulatory process • Growing appeal of ENERGY STAR labeling to consumers • Reduction in capital gains tax

In thinking about a vision for the windows industry, participants considered a 20-year time horizon. The 20-year horizon stimulates industry members to imagine their ideal world without concern for present-day barriers. Executive forum participants developed their vision of the future using graphical facilitation techniques. Facilitators asked two groups of participants to imagine future cover stories that heralded their success. They envisioned that, in the next 20 years, the U.S. window industry will offer its customers imaginative new products that challenge traditional perceptions. Windows will become active, integral parts of building climate, energy, information, and structural systems. Responsible manufacturing practices, material selection, and energy efficiency characteristics will combine to also make windows a solution to environmental concerns. To help customers understand the added value that windows offer them over competing building products, members of the window industry will become premier educators. All these efforts will increase demand for windows as an alternative to competing building components and appliances, thereby enhancing the industry's growth and contributing to its strength.

Figure 5-1. Demonstration of Advanced Window Technology at NREL

Industry members condensed their vision into the compelling vision statement below.

Vision Statement

In 2020, consumers recognize windows[1] as affordable "appliances in the wall" that are active and interactive parts of a true building system. Windows offer added value by providing energy, entertainment, and information with enhanced comfort, lighting, security, and aesthetics, in harmony with the natural environment.

The vision statement is supported by six vision elements as articulated in the Executive Forum discussion. These are:

1. Windows as an integral part of a building "system"
2. Active, smart glass and windows

3. Informed consumers at all levels
4. More glass and windows used in buildings
5. Windows as an environmental solution
6. Windows as an energy source

Following the Executive Vision Forum, the window industry organized a Window Technology Roadmap Workshop, held in January 1999 in Leesburg, Virginia. Over 30 representatives from the fenestration industry, government, environmental organizations, and research groups met to complete an industry-wide plan for achieving the industry vision. This collaborative workshop helped identify key targets of opportunity, technology barriers, and research priorities to meet the vision.

This workshop was followed up with a survey of workshop participants. The survey questionnaire was distributed to participants and asked respondents to identify specific research needs an rate the investment required, potential contribution toward the vision, and the certainty of success for each. This roadmap represents the aggregation of those responses.

BARRIERS AND STRATEGIES TO OVERCOME THEM

In order to achieve the vision, the window industry must overcome barriers. During the Leesburg, Virginia, meeting, members identified the key barriers in the areas of technology, market, and policy. In addition, participants outlined strategies for overcoming these barriers. Participants voted on which barriers to discuss further during the workshop, and specific actions were developed to address the most important barriers. This germinal roadmap emphasized industry's near-term priorities in each of the three areas: technology, market and policy.

Technology Barriers

The two most important technology barriers were identified to be: (1) the lack of integration tools and forms to achieve true system integration; and (2) an ambiguous definition of "durability" and its implications for warranty. Table 5-2 shows a list of all the barriers that were identified. Table 5-3 presents the technical barriers that were selected for further discussion at the Leesburg workshop. These tables also include a summary of the strategic actions that can be taken to overcome the identified barriers.

Table 5-2. Technology Barriers

- Lack of integration tools and forms to achieve true system integration
- Ambiguous definition of "durability" and its implications for warranty
- High cost of manufacturing, materials, and research
- Consumer and corporate mindset against the vision
- Absence of interconnection and control technologies for building systems
- Presence of competing technologies such as opaque walls and artificial lighting
- Long product development and cycle times

Table 5-3. Actions for Selected Technology Barriers

Barrier	Action
Lack of integration tools and forms	• Define interface standards and protocols for integrating different building system components. • Develop strategies and hardware necessary to optimize integrated building systems. • Define performance metrics for comfort, system integration, energy, cost, and environmental impacts. • Develop methods for measuring the value of integrated systems.
Definition of durability and implications for warranty	• Establish a system for rating products on the basis of durability. • Define appropriate durability and warranty periods for different window components. • Develop products that encourage consumers to upgrade as features advance (replaceable, portable, modular, high value).

Other actions related to technology were identified as "high-priority" and were broken down by near-term (0-3 years), mid-term (3-10 years), and long-term (10-20 years). These actions are shown in Table 5-4.

Table 5-4. High Priority Technology Action Items

Near-Term (0-3 years)	• Define standards and protocols for integrating different building components. • Develop strategies and hardware necessary to optimize integrated building systems. • Define performance metrics for comfort, system integration, energy, cost, and environmental impacts. • Develop methods for measuring the value of integrated systems. • Establish a system for rating products on the basis of durability. • Define appropriate durability and warranty periods for different window components.
Mid-Term (3-10 years)	• Develop analytical tools to assist manufacturers in designing and marketing efficient windows. • Develop methods to measure and prove durability of fenestration products. • Support, specify, and identify applications for improved technology, including breakthrough materials and manufacturing processes. • Develop products that encourage consumers to upgrade as features advance (replaceable, portable, modular, high value).

(Continued)

Table 5-4 (*Cont'd*)

Mid-Term (3-10 years)	• Establish a regionally sensitive national building code.
Long-Term (10-20 years)	• Develop long-term photovoltaic products that can be integrated in fenestration products. • Develop superior insulating materials and components for fenestration products. • Develop integrated electronics in fenestration products.

Market Barriers

The two most important market barriers were identified to be: (1) lack of educated demand; and (2) high first costs of products. Other market barriers are shown in Table 5-5. Specific action items for the top two barriers are presented in Table 5-6.

Table 5-5. Market Barriers

- Lack of educated demand
- High first cost
- Fragmentation in the fenestration and building industries
- Lack of product differentiation by non-cost attributes
- Resistance to partnering among industry members

Table 5-6. Actions for Selected Market Barriers

Barrier	*Action*
Lack of educated demand	• Understand the market by clearly identifying the audience. • Create and use tools.

(*Continued*)

Table 5-6 (*Cont'd*)

Lack of educated demand	• Understand current technology and potential applications and specify technology needs as identified by user expectations. • Establish partnerships through collaborative work between multiple stakeholders and resource groups.
High first cost	• Educate stakeholders and end users on true long-term cost benefits. • Conduct a value-based market analysis. • Support, specify, and identify applications for improved technology, including breakthrough materials and manufacturing processes. • Provide incentives such as financing programs and low interest loans, perhaps as an expanded ENERGY STAR component.

Other market action items were identified as "high-priority." These are shown in Table 5-7.

Table 5-7. High Priority Market Action Items

Near-Term (0-3 years)	• Establish partnerships through collaborative work among multiple stakeholders and resource groups. • Conduct a value-based market analysis. • Support, specify, and identify applications for improved technology, includ

(*Continued*)

Table 5-7 (*Cont'd*)

ing breakthrough materials and manufacturing processes.
- Provide incentives such as financing
programs and low interest loans,
perhaps as an expanded Energy Star
component.

Policy Barriers

Although a number of policy barriers were identified (see Table 5-8), only one item was selected for consideration of strategic actions. This item was related to poorly enforced and inconsistent building codes that contradict Department of Energy and industry goals. The actions related to this barrier are shown in Table 5-9.

Table 5-8. Policy Barriers

- Dissimilar, poorly enforced, and inconsistent building codes that contradict DOE and industry goals
- No teeth in code enforcement
- Undermining of local code enforcement by special interest groups
- Limited Congressional support for end-use versus supply-side programs
- Lack of compelling national energy policy
- Lack of Congressional support for integrated roadmaps
- Lack of clarity about how to measure success

Table 5-9. Actions for Selected Policy Barriers

Barrier	*Action*
Dissimilar, poorly enforced, and inconsistent	• Combine the three existing codes by supporting the ICC, professional lobbying, or creating a core industry group.

(*Continued*)

Table 5-9 (*Cont'd*)

Dissimilar, poorly enforced, and inconsistent building codes	• Develop recommendation for Congressional legislation on establishing a regionally sensitive national building code. • Develop communication channels among building industry groups to address integration issues in areas of education, research, and collaboration.

Other policy action items identified as "high priority" include those identified in Table 5-10.

Table 5-10. High Priority Policy Action Items

Near-Term (0-3 years)	• Combine the three existing codes by supporting the International Code Council (ICC), professional lobbying, or creating a core industry group. • Educate local building inspectors. • Develop communication channels among building industry groups to address integration issues in areas of education, research, and collaboration.
Mid-Term (3-10 years)	• Establish a regionally sensitive national building code

MAPPING RESEARCH NEEDS AND STRATEGIES

The two workshops stimulated creative thinking and developed general consensus about the future of the window industry. They also identified interesting market transformation activities needed to support the vision. However, the workshops did not identify research needs and strategies in enough detail to complete the technology roadmap. Time constraints were partially the cause, and participants hesitated to discuss detailed research ideas in front of their competitors, even if the ideas were precompetitive.

To collect the necessary technical information free of the limitations imposed by the workshop environment, DOE distributed a questionnaire to over 20 workshop participants and researchers. The surveys asked respondents to identify and describe specific research needs in the vision's five technical elements:

- Building integration—structural, power, and data interconnection between the window and the rest of the building
- Information display—passive, active, or interactive display of text or images
- Energy supply and conservation—annual or, ideally, instantaneous net provider of energy to the building
- Environmental harmony—minimal negative environmental impacts over the product life cycle
- Enhanced traditional features—improved window characteristic

For each research need, respondents also rated the investment required, the potential contribution toward each element of the vision, and the certainty of success. The respondents were contacted by phone to clarify and further develop their responses.

Respondents identified 65 unique research activities that could move the industry toward its vision by overcoming technical barriers. In alphabetical order, they are:

- ADVANCED HOLOGRAMS-Produce holographic images on windows

- AEROGELS-Incorporate non-opaque, highly insulating aerogel into insulating glass units

- ALTERNATIVE GLAZING MATERIALS-Develop more durable and efficient glazing materials

- ALTITUDE ADAPTIVE IG-Redesign IG units to eliminate breakage due to bulging at high altitude

- BILLET STOCK FROM RECYCLE-Develop suitable process for making billet out of recycled aluminum

- BLAST-RESISTANT WINDOWS-Develop new, cost-effective, architecturally acceptable blast-resistant window materials

COLORADO COLLEGE LIBRARY
COLORADO SPRINGS
COLORADO

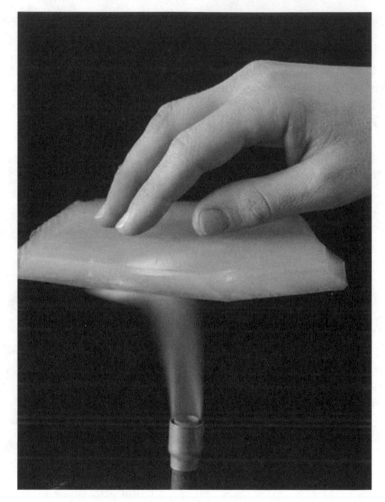

Figure 5-2. Aerogel Material Providing Better Insulation Properties

- BUILDING ENERGY SOFTWARE-Develop software to predict the energy performance of a building

- BUILDING INTEGRATION DEMONSTRATION-Demonstrate an integrated building system with windows

- COATING EQUIPMENT-Design coating equipment flexible enough to apply a variety of coatings

- COLOR PHOTOCHROMICS-Expand the color availability of photochromic materials

- DAYLIGHTING RATING-Provide a rating to measure the amount of daylighting provided by a window

- ELECTROCHROMIC DISPLAY-Develop "smart" windows

- ELECTROCHROMIC FAILURE MODES-Identify electrochromic failure modes

- ELECTROCHROMIC SCALE-UP-Prove electrochromics in commercial window sizes

- ELECTROCHROMIC SERVICE-LIFE PREDICTION-Develop models to predict service life of electrochromics based on product specs and tests

- ENERGY-EFFICIENT EXTRUSION-Reduce energy intensity of aluminum extrusion

- ENVIRONMENTALLY BENIGN PHOTOVOLTAICS-Research and utilize environmentally benign photovoltaic (PV) materials

- EXTERIOR DISPLAY-Display images on window exteriors

- FENESTRATION DURABILITY-Research materials and finishes to extend efficient fenestration life

- FIRE-RATED WINDOWS-Develop lower-cost alternative materials for fire-rated windows

- GAS RETENTION-Test and predict gas concentration in IG units

- GLASS/FRAME RATIO-Increase vision area without a corresponding increase in framing

- HIGH-SECURITY WINDOWS-Develop new, stronger, cost-effective, architecturally compatible materials for high security

- HOLOGRAMS-Exploit holography to direct exterior lighting within the interior space

- HOLOGRAPHIC MODELING-Improve the modeling of the transmission of sunlight through holograms

- IDENTIFY MARKETS FOR PROCESS WASTE-Find partners to use waste streams from window manufacturing operations

- INSULATING COATINGS-Develop new colored architectural coatings that reduce conductive heat loss through window frames and sashes

- INSULATING COMPONENTS-Develop new alloys or composites that reduce conductive heat loss through window components

- INTEGRAL SMART SYSTEMS-Develop self-contained power supplies, sensors, controllers, and actuators to actively control heat and light transmission through the window

- INTEGRAL WIND POWER RECOVERY-Integrate components into windows to capture wind energy

- INTEGRAL WIRING-Incorporate wiring or wiring runs into the window

- INTEGRATED BUILDING ENERGY SYSTEM SOFTWARE-Develop low-cost, user-friendly software to assess the energy savings inherent in integrated building systems

- INTERIOR DISPLAY-Display images on window interiors

- INTERIOR LIGHTING SOURCE-Transmit light from spandrel through ceiling space

- INTERIOR PASSIVE LIGHTING-Develop light shelves for curtain wall and window wall applications

- LARGER PV PANELS-Produce photovoltaic panels in sizes larger than 2'x4'

- LASER IMPRINTING-Improve laser imprinting process for holograms on a commercial scale

- LOW-COST IG-Develop new ways to produce affordable IG units

- LOW-E COATINGS-Develop new generation of scratch-resistant, cleanable coating materials

- MODULAR WINDOWS-Design new window system with permanent frames and modular windows

- MONOCHROMIC ELECTROCHROMIC DISPLAY-Electrochromic display

- MONOLITHIC TRANSPARENT INSULATING MATERIALS-Develop new non-glass insulating materials

- MULTICHROMIC ELECTROCHROMIC DISPLAY-Electrochromic color display

- PHOTOCHROMIC SCALE-UP-Prove photochromics in commercial window sizes

- POWER SUPPLY MINIATURIZATION-Develop miniature, self-contained power supplies for active windows

- POWER SYSTEM BALANCING-Develop power balancing/conditioning components that are integral to the window

- PROJECTED DISPLAY-Project images onto windows similar to a "heads up" display

- PROTOCOL FOR COMMUNICATION-Develop a means to communicate between various electronic components

- PV COATINGS-Develop photovoltaic coatings

- PV PANEL COLORS-Expand the color availability of photovoltaic panels

- PV THIN FILM-Incorporate thin-film photovoltaics into fenestration products

- PV VISION GLASS-Develop semitransparent photo-voltaic glazing

- RECYCLABILITY-Improve ability to disassemble dissimilar window materials for recycling

- SLOPE U-FACTOR-Develop a U-factor rating suited to sloped skylights

- SMART PHOTOCHROMICS-Develop photochromic glazings that also regulate heat transmission

- SOFTWARE TOOLS TO QUANTIFY PERFORMANCE-Provide a simpler means to quantify performance through use of software

- SOLAR HEAT GAIN-Develop a solar heat gain rating suited to skylights

- STRONGER SEALANT-Strengthen the sealant bond in structural windows

- SUNSCREENING-Develop skylight accessories to control conductive and radiant heat transmission

- THERMAL MODELING SOFTWARE-Continue to improve the usability, flexibility, and cost of 3-D thermal models of window systems

- THERMOCHROMICS SCALE-UP-Prove thermochromics in a commercial size window

- UV RESEARCH BY MEDICAL RESEARCHERS-Research to understand the effects of ultraviolet light on humans

- VACUUM GLASS-Develop commercially viable vacuum glass

- VENTILATION-Develop fenestration systems that regulate or condition outdoor air for indoor use

- WINDOW SELECTION SOFTWARE-Develop software to select windows based on impacts on building energy consumption

These research activities were grouped into the following eight research areas:

1. Imaging—display of images or text on the window surface

2. Energy production and supply—development of window-based photovoltaic materials

3. Light transmission—control of radiant light and heat transmission through windows

4. Insulation—control of heat conduction through windows

5. Analytical tools—modeling of window-related phenomena and development of software-based tools

6. Manufacturing—equipment and processes for producing windows and window-related components

7. Design—design of buildings and building systems including windows

8. Electronics—development of integral components for controlling and powering window features

Research areas clarify the extent to which types of research needs contribute to various vision elements. This clarification can help an organization fund or organize efforts in the research areas that best support those vision elements it finds most appealing. For example, DOE may decide to emphasize research in those areas that best contribute to the vision's energy element. For organizations that conduct research, research areas can help them decide how they might best contribute to the vision based on how well their competencies match each element.

Table 5-11 presents a map for each of the priority research areas by topic.

Figure 5-3. PV Integrated Skylight at Presidio National Park

Table 5-11. Research Priorities by Research Area

Research Area	Continuing Research	Future Research
Imaging	• Projected display • Interior display	• Electrochromic display • Advanced holograms • Exterior display • Monochromic display • Multichromic display
Energy Production and Supply	• Larger PV panels • PV vision glass • PV thin film	• Environmentally benign • PV materials • PV coatings • PV panel colors • Integral wind power
Light Transmission	• Electrochomics scale-up • Photochromics scale-up	• Smart photochromics • Color photochromics

(Continued)

Table 5-11 (*Cont'd*)

Light Transmission	• Themochromics scale-up • Holograms • Low-e coatings • UV research by medical researchers • Interior lighting source	• Daylighting rating
Insulation	• Insulating components • Aerogels • Monolithic transparent insulating materials • Vacuum glass • Gas retention	• Insulating coatings • Alternative glazing materials
Analytical Tools	• Thermal modeling • Building energy software • Solar heat gain • Slope U-factor • Holographic modeling • EC failure modes • Life-cycle software/analysis • Window selection software	• Tools to quantify performance • EC service-life prediction
Manufacturing	• Billet stock from recycle • Energy-efficient extrusion • Laser imprinting • Low cost of efficient IG • Coating equipment • Markets for process waste	• Recyclability
Design	• Altitude adaptive IG • Stronger sealant • High-security windows • Glass/frame ratio	• Modular windows • Ventilation

(Continued)

Table 5-11 (*Cont'd*)

Design	• Blast-resistant windows • Fenestration durability • Fire-rated windows • Sunscreening • Interior passive lighting • Building integration demonstration	

Electronics	• Power supply
	  miniaturization
	• Integral wiring
	• Power system balancing
		• Integral smart system
		• Protocol for
		  communication

CONCLUSION:
NEXT STEPS OF THE WINDOWS ROADMAP

Although product development is essential to the long-term success of the industry, it is a primary basis for competition among companies and is best left to the individual efforts of company proprietary research and development programs. However, studies of the fundamental physical characteristics of windows and complementing technologies are needed. Individual company researchers and product developers should use the results of this fundamental research to advance proprietary product development and to promote competition.

Achieving the goals identified in this roadmap will require collaboration with government and other industries to leverage research and development funds. Collaboration will be required for the following long-term research objectives:

• Develop long-term photovoltaic products that can be integrated in fenestration products.

• Develop superior insulating materials and components for fenestration products.

• Develop analytical tools to assist manufacturers in designing and marketing efficient windows.

- Develop methods to measure and prove durability of fenestration products.

- Develop integrated electronics in fenestration products.

- Support, specify, and identify applications for improved technology, including breakthrough materials and manufacturing processes.

- Develop products that encourage consumers to upgrade as features advance (replaceable, portable, modular, high value).

However, research and development alone will not lead to achieving the vision. Government and industry will need to continue working together to address the market and policy barriers facing the window industry. Objectives include:

- Define interface standards and protocols for integrating different building system components.

- Develop communication channels among building industry groups to address integration issues in areas of education, research, and collaboration.

- Develop strategies and hardware necessary to optimize integrated building systems.

- Define performance metrics for comfort, system integration, energy, cost, and environmental impacts.

- Develop methods for measuring the value of integrated systems.

- Establish a system for rating products on the basis of durability.

- Define appropriate durability and warranty periods for different window components.

- Understand current technology and potential applications and specify technology needs as identified by user expectations.

- Educate stakeholders and end users on true long-term cost benefits.

- Provide incentives such as financing programs and low-interest loans, perhaps as an expanded ENERGY STAR component.

Several next steps are needed to implement the vision and roadmap and to pursue research opportunities. First, an industry task group should be formed to address appropriate industry and government roles in implementing the roadmap. This would include the establishment of ad hoc working groups to examine the eight research areas in more depth and develop detailed research plans for each area. Simultaneously, DOE should identify areas in the roadmap that coincide with beneficial public policy and align its federal research agenda accordingly. And finally, industry and government should work together to continue "course correction" meetings to ensure that the roadmap is a living, evolving document.

[**Editor's Note:** This chapter represents an abridged and edited version of the U.S. Department of Energy report entitled *Window Industry Technology Roadmap* published by the Office of Building Technologies, U.S. DOE, April 2000, DOE/ GO-102000-0980. The editor is grateful to the U.S. Department of Energy for permission to publish an abridged version of this report.]

Reference
[1]The term "window" in the vision statement, as well as in the roadmap itself, refers to fenestration products, including windows, doors, and skylights.

Chapter 6

Hydrogen Energy Technology Roadmap

A Glimpse of the Future

By 2020...

- Hydrogen will be used in refrigerator-size fuel cell units to produce electricity and heat for the home.
- Vehicles that operate by burning hydrogen or by employing hydrogen fuel cells will be commercially viable and emit nothing more than water out of their tailpipes.
- Hydrogen refueling stations that use natural gas to produce hydrogen will be available in urban areas to refuel hydrogen vehicles.
- Micro-fuel cells that use small tanks of hydrogen will be operating everything from mobile generators to electric bicycles to vacuum cleaners.
- Large 250 kW stationary fuel cells, alone or in tandem, will be used for backup power and as a source of distributed generation supplying electricity to the utility grid.

INTRODUCTION

There are many benefits that could be expected from a future "hydrogen economy." The expanded use of hydrogen as an energy carrier over the next 20 years could help address national concerns about energy security, global climate change, and air quality. Because hydrogen can be derived from a variety of domestically produced primary sources (including fossil fuels, renewables, and nuclear power), the country could reduce its dependence on foreign sources of energy. In addition, the by-products of hydrogen conversion are generally benign for human health and the environment.

Despite these benefits, realization of a hydrogen economy faces multiple challenges. Unlike gasoline and natural gas, hydrogen has no existing, large-scale supporting infrastructure—and building one will require major investment. Although hydrogen production, storage, and delivery technologies are currently in commercial use by the chemical and refining industries, existing hydrogen storage and conversion technologies are still too costly for widespread use in energy applications. Lastly, existing energy policies do not promote consideration of the external environmental and security costs of energy that would encourage wider use of hydrogen.

This chapter presents a *hydrogen technology roadmap* that outlines the necessary technological, market, and policy measures needed to move from a fossil fuel economy to a hydrogen economy. The roadmap identifies a vision for future hydrogen use in the U.S., the current status of hydrogen technologies, barriers to achieving the hydrogen economy, and strategies to help overcome these barriers. Thus, this roadmap provides a plan of action for policy-makers and industry to move towards a hydrogen economy.

THE ROADMAP PROCESS

A Hydrogen Vision

The development of hydrogen as a future energy option will necessitate an unprecedented level of sustained and coordinated activities by diverse stakeholders. Recognizing the need to develop a coordinated national agenda, the U.S. Department of Energy (DOE) initiated a national hydrogen vision and roadmap process to incorporate the opinions and viewpoints of a broad cross-section of those stakeholders. The process involved two key meetings: the National Hydrogen Vision Meeting and the National Hydrogen Energy Roadmap Workshop.

The National Hydrogen Vision Meeting was held on November 15-16, 2001, in Washington, DC. Participants included more than 50 business executives and public policy leaders from federal and state agencies, the U.S. Congress, and environmental organizations. The U.S. DOE initiated the meeting in response to recommendations in the National Energy Policy regarding hydrogen technologies. The aims of the meeting were to identify a common vision for the hydrogen economy, the time frame in which such a vision could be expected to occur, and the key milestones for achieving it.[1]

Vision for the Hydrogen Economy

Hydrogen is America's clean energy choice. Hydrogen is flexible, affordable, safe, domestically produced, used in all sectors of the economy, and in all regions of the country.

Major findings from the vision meeting include the following:

- Hydrogen energy could play an increasingly important role in America's energy future, as it has the potential to help reduce dependence on petroleum imports and lower pollution and greenhouse gas emissions.
- The transition to a hydrogen economy has begun, and could take several decades to achieve.
- The development of hydrogen technologies needs to be accelerated.
- There are "chicken-and-egg" issues regarding market segment development and how supply and demand will push or pull these activities.
- Federal and state governments will need to implement and sustain consistent energy policies that elevate hydrogen as a priority.

The vision meeting provided a starting point for stakeholders to begin thinking about the future of hydrogen use in the United States. From this vision meeting, the hydrogen roadmap was born.

The Hydrogen Roadmap

After the Hydrogen Vision Meeting concluded, participants engaged in the National Hydrogen Energy Roadmap Workshop (held April 2-3, 2002) in Washington, DC. Approximately 220 technical experts and industry practitioners from public and private organizations participated in the meeting. Seven leaders from industry and academia with expertise in hydrogen systems helped guide the roadmap development process.

Each of these leaders was assigned to a particular segment of the hydrogen system. Table 6-1 identifies the leaders, their affiliation, and the segment to which they were assigned. During the workshop, participants discussed key needs that should be addressed in order to achieve the hydrogen vision. The remainder of this chapter discusses each of the roadmap segments, as shown in the last column of Table 6-1.

Table 6-1. Hydrogen Roadmap Leaders

Leader	Affiliation	Roadmap Segment
Frank Balog	Ford Motor Company	Applications
Mike Davis	Avista Labs	Energy Conversion
Art Katsaros	Air Products and Chemicals, Inc.	Delivery
Gene Nemanich	Chevron Texaco Technology Ventures	Production
Alan Niedzwiecki	Quantum Technologies	Storage
Joan Ogden	Princeton University	Systems Integration
Jeff Serfass	National Hydrogen Association	Public Education and Outreach

THE HYDROGEN SYSTEM: A STATUS REPORT

Hydrogen can be produced in centralized facilities or at decentralized locations where it will be used onsite. From centralized facilities, it is distributed to an energy conversion device via pipeline, or stored and shipped via rail or truck. When produced onsite, hydrogen can be stored and/or fed directly into conversion devices for stationary, mobile, and portable applications.

Figure 6-1 shows the hydrogen production, delivery, storage, conversion, and application system. Each one of these system components will be addressed individually in the following sections. For each system component, we provide a status report, explain a vision for the future, identify barriers, and highlight strategies for overcoming those barriers.

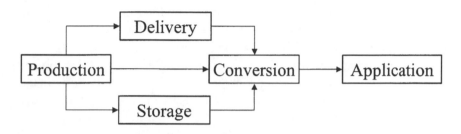

Figure 6-1. Hydrogen System Overview

HYDROGEN PRODUCTION

Production Status

Hydrogen can be produced from a variety of sources, including: fossil fuels, renewable sources (e.g., wind, solar, biomass), nuclear power, solar thermochemical reactions, and solar photolysis, to name a few.

The U.S. hydrogen industry currently produces nine million tons of hydrogen per year for use in chemical production, petroleum refining, metals treating, and electrical applications. This is enough hydrogen to fuel 20-30 million hydrogen-fueled cars annually. However, hydrogen now is primarily used as a feedstock, intermediate chemical, or, on a much smaller scale, a specialty chemical. Only a small portion of the hydrogen produced today is used as an energy carrier.

Although hydrogen is the most abundant element in the universe, it does not naturally exist in large quantities or high concentrations on Earth—it must be produced from other compounds such as water, biomass, or fossil fuels. Various methods of production have unique needs in terms of energy sources (e.g., heat, light, electricity) and generate unique by-products or emissions.

Steam methane reforming accounts for 95% of the hydrogen produced in the United States. This is a catalytic process that involves reacting natural gas or other light hydrocarbons with steam to produce a mixture of hydrogen and carbon dioxide. The mixture is then separated to produce high-purity hydrogen. This method is the most energy-efficient commercialized technology currently available.

Partial oxidation of fossil fuels in large gasifiers is another method of thermal hydrogen production. This involves the reaction of a fuel with a limited supply of oxygen to produce a hydrogen mixture, which is then purified. Partial oxidation can be applied to a wide range of hydrocarbon feedstocks, including natural gas, heavy oils, solid biomass, and coal. Its primary by-product is carbon dioxide.

Hydrogen can also be produced by using electricity in electrolyzers to extract hydrogen from water. Currently this method is not as efficient or cost effective as using fossil fuels in steam methane reforming and partial oxidation, but it would allow for more distributed hydrogen generation and open possibilities for using electricity made from renewable and nuclear resources.

Other methods hold the promise of producing hydrogen without carbon dioxide emissions, but all of these are still in early development phases. They include thermochemical water-splitting using nuclear or solar heat, photolytic (solar) processes using solid state techniques, fossil fuel hydrogen production with carbon sequestration, and biological tech-

Figure 6-2. Steam Reforming Facility Producing Hydrogen

niques (algae and bacteria) that generate hydrogen from hydrogen containing materials.

Production Vision

The vision for hydrogen production has been identified as follows:

Hydrogen will become a premier energy carrier, reducing U.S. dependence on imported petroleum, diversifying energy sources, and reducing pollution and greenhouse gas emissions. It will be produced in large refineries in industrial areas, power parks and fueling stations in communities, distributed facilities in rural areas, and on-site at customers' premises. Thermal, electric, and photolytic processes will use fossil fuels, biomass, or water as feedstocks and release little or not carbon dioxide into the atmosphere.

A pathway for scaling up hydrogen use would build from the existing hydrogen industry. To foster the initial growth of distributed mar-

Figure 6-3. Photobiological Hydrogen Production Facility at NREL

kets, small reformers and electrolyzers will provide hydrogen for small fleets of fuel cell-powered vehicles and distributed power supply. The next stage of development will include mid-sized community systems and large, centralized hydrogen production facilities with fully developed truck delivery systems for short distances and pipeline delivery for longer distances. As markets grow, costs will drop through economies of scale and technological advances; carbon emissions will decrease with commercialization of carbon capture, sequestration, and advanced conversion methods using photolytic, renewable, and nuclear technologies.

Production Barriers

Multiple challenges must be overcome to achieve the vision of secure, abundant, inexpensive, and clean hydrogen production with low carbon emissions. Barriers that have been identified include:

- **Hydrogen production costs are high relative to conventional fuels**. With most hydrogen currently produced from hydrocarbons, the cost per unit of energy delivered through hydrogen is higher than the cost of the same unit of energy from the hydrocarbon itself.

- **Low demand inhibits development of production capacity**. Although there is a healthy, growing market for hydrogen in refineries and chemical plants, there is little demand for hydrogen as an energy carrier. Without demand, there is little incentive for industry to completely develop, optimize, and implement existing and new technologies.

- **Current technologies produce large quantities of carbon dioxide and are not optimized for making hydrogen as an energy carrier**. Existing production technologies can produce vast amounts of hydrogen from hydrocarbons but emit large amounts of carbon dioxide into the atmosphere. Technical improvements are needed to reduce costs, improve efficiencies, and produce hydrogen with little or not carbon emissions.

- **Lack of demonstration sites**. Stakeholders need a basic understanding of the different sources of hydrogen production before they will be willing to embrace the concepts. Demonstrations are needed to gain confidence in these technologies.

Production Strategies and Paths Forward

The specific needs and actions required to address the above barriers differ for each hydrogen production technology. No single technology meets all of the criteria of the production vision. Some suggested strategies are below.

- **Enact policies that foster both technology and market development**. Government support for research and development should focus on developing advanced renewable and low-carbon-emitting methods plus carbon dioxide capture and sequestration technologies.

- **Improve gas separation and purification processes**. Lowering the cost of multi-fuel gasifiers and developing low-cost, high efficiency methods for hydrogen purification will help lower costs of hydrogen production, especially at decentralized sites.

- **Develop and demonstrate small reformers**. Small reformers that run on natural gas, propane, methanol or diesel can provide hydrogen to some of the first fleets and retail sales points, reducing overall costs.

- **Optimize and reduce costs of electrolyzers**. Efforts to improve the efficiency and lower the costs of electrolyzers are important. Although electrolysis is currently more expensive than thermal production, a better understanding of high-temperature and high-pressure electrolysis could bring costs down.

- **Develop advanced renewable energy methods for hydrogen production**. Semiconductors that lower costs and improve efficiencies of photolytic processes to split water and produce hydrogen are needed. Biological systems should also be developed as a potential "low-tech" way to produce hydrogen.

- **Develop advanced nuclear energy methods to produce hydrogen**. Research is needed to identify and develop methods for economically producing hydrogen with nuclear energy, which would avoid carbon emissions. Thermochemical water splitting using high temperature heat from advanced nuclear reactors could be included in future nuclear plant designs.

- **Develop methods for large-scale carbon dioxide capture and sequestration**. A cost-effective way to capture and sequester carbon dioxide would facilitate the production of vast quantities of hydrogen with low carbon emissions. Capture systems would need to be engineered into plant designs for steam methane reformers and multi-fuel gasifiers to lower the overall systems costs.

- **Demonstrate production technologies in tandem with applications**. Demonstrations are expensive, especially since there may be little initial demand for the hydrogen produced. Demonstrations that integrate production technology with other elements of the hydrogen infrastructure, including a market use, will be more cost effective.

In sum, research, development, and demonstrations are needed to improve and expand methods of economically producing hydrogen. Production costs need to be lowered, efficiency improved, and carbon sequestration techniques developed. Better techniques are needed for both central-station and distributed hydrogen production. Efforts should focus on existing commercial processes such as steam methane reforming, multi-fuel gasifiers, and electrolyzers, and on the development of advanced techniques such as nuclear thermochemical water splitting, photoelectrochemical electrolysis, and biological methods.

HYDROGEN DELIVERY

Delivery Status
A key element of the overall hydrogen energy infrastructure is the delivery system that moves the hydrogen from its point of production to an end-use device. Delivery system requirements necessarily vary with the production method and end-use application. At present, hydrogen is currently transported from a limited number of production plants by pipeline or by road via cylinders, tube trailers, and cryogenic tankers, with a small amount shipped by rail car or barge.

Pipelines are employed as an efficient means to supply customer needs. The pipelines are currently limited to a few areas of the U.S. where large hydrogen refineries and chemical plants are concentrated, such as in Indiana, California, Texas, and Louisiana. The pipelines are owned and operated by merchant hydrogen producers.

Hydrogen distribution via high-pressure cylinders and tube trailer has a range of 100-200 miles from the production or distribution facility. For long-distance distribution of up to 1000 miles, hydrogen is usually transported as a liquid in super-insulated, cryogenic, over-the-road tankers, railcars, and barges and is then vaporized for use at the customer site.

Delivery Vision

The vision for hydrogen delivery was identified by stakeholders as follows:

> A national supply network will evolve from the existing fossil fuel-based infrastructure to accommodate both centralized and decentralized production facilities. Pipelines will distribute hydrogen to high-demand areas, and trucks and rail will distribute hydrogen to rural and other lower-demand areas. On-site hydrogen production and distribution facilities will be built where demand is high enough to sustain maintenance of the technologies.

Delivery Barriers

A comprehensive delivery infrastructure for hydrogen faces numerous scientific, engineering, environmental, institutional, and market challenges, including:

- **An economic strategy is required for the transition to a hydrogen delivery system**. Since fueling economics depend on volume, the chicken-and-egg dilemma (which comes first: fuel or end use applications?) impedes the installation of an effective infrastructure. There is no simple reconciliation between the level of investments required to achieve low costs and the gradual development of the market.

- **Full life-cycle costing has not been applied to delivery alternatives**. Any strategy to select an appropriate delivery system should involve full life-cycle costing of the options. This includes full social costing of alternatives.

- **Hydrogen delivery technologies cost more than conventional fuel delivery**. The high cost of hydrogen delivery methods could lead to the use of conventional fuels and associated delivery infrastructure up to the point of use, and small-scale conversion systems to make hydrogen onsite. However, cost effective means do not currently exist to generate hydrogen in small-scale systems.

- **Current dispensing systems are inconvenient and expensive**. Customers expect the same degree of convenience, cost performance, and safety when dispensing hydrogen fuel as when dispensing conventional fuels. Current hydrogen fueling solutions and designs are not sufficiently mature to provide this convenience.

- **Codes and standards are not developed**. There is currently a lack of codes and standards for hydrogen delivery and a lack of harmonization between national and international codes.

Delivery Strategies

Current delivery systems will need to expand significantly to deliver hydrogen to all regions of the country in a safe and affordable manner. Distributed hydrogen production is likely to play a significant role, but alternative delivery systems tailored to consumer applications (such as the transport of hydrogen in safe, solid metal alloy hydrides, carbon nanomaterials, and other chemical forms) need to be developed to transport hydrogen to end-use sites on an as-needed basis. Some strategies to meet the delivery vision include:

- **Develop a demonstration rollout plan**. A hydrogen delivery infrastructure needs to be started in several regions of the United States. Government-sponsored pilot testing of refueling systems would help establish a basis for certifying components of fuel stations.

- **Develop a consensus view on total costs of delivery alternatives**. Analysis of the total costs of delivery alternatives needs to be conducted. Analysis should weigh options that address all potential fuel delivery points, the cost of maintaining existing fuel infrastructure, and the suitability of the existing infrastructure for future hydrogen use.

- **Increase research and development on delivery systems**. Improvements are needed in areas such as hydrogen detectors; odorization;

materials selection for pipelines, seals and valves; and transportation containers for hydrogen. Technology validation should address research and development needs for fueling components such as high-pressure, breakaway hoses; hydrogen sensors; compressors; on-site hydrogen generation systems; and robotic fuelers.

- **Create testing and validation protocols**. Testing and validation should be ongoing. An organization should be established to perform testing and certification and to identify components that require validation and testing protocols.

HYDROGEN STORAGE

Storage Status

Storage issues cut across the production, transport, delivery, and end-use application of hydrogen as an energy carrier. Mobile applications are driving the development of safe, space-efficient, and cost-effective hydrogen storage systems, yet other applications will benefit substantially from technological advances made for on-vehicle storage systems.

Currently, hydrogen can be stored as a gas or liquid or in a chemical compound. The storage of compressed hydrogen gas in tanks is the most mature technology, though the very low energy density of hydrogen means compression between 5,000 and 10,000 psi is needed to improve vehicle range. Liquid hydrogen takes up less storage space, but requires cryogenic containers. In addition, liquefaction of hydrogen is an energy-intensive process.

More recently, developments in metal hydrides or carbon nanotubes have shown promise as a hydrogen storage device. When hydrogen is needed, it can be released from these materials under certain temperature and pressure conditions. There are also chemical hydrides being investigated that bind hydrogen in a chemical compound and then release that hydrogen through a catalyzed chemical process. However, these novel methods are currently expensive. Table 6-2 identifies a number of hydrogen storage alternatives.

No current technology satisfies all the desired storage criteria, which include: low cost, low weight, low volume, safe, easy-to-handle, long-term storage viability, high storage efficiency.

Table 6-2. Hydrogen Storage Alternatives

Compressed Fuel Storage	• Cylindrical tanks • Quasi-conformable tanks
Liquid Hydrogen Storage	• Cylindrical tanks • Elliptical tanks • Cryotanks • High-pressure liquid tanks
Solid State Conformable Storage	• Hydride materials • Carbon adsorption
Chemical Hydrides	• Alkaline liquids

Storage Vision

Stakeholders identified the vision for hydrogen storage as follows:

A selection of relatively lightweight, low-cost, and low-volume hydrogen storage devices will be available to meet a variety of energy needs. Pocket-sized containers will provide hydrogen for portable telecommunications and computer equipment, small and medium hydrogen containers will be available for vehicles and on-site power systems, and industrial sized storage devices will be available for power parks and utility-scale systems. Solid-state storage media that use metal hydrides will be in mass production as a mature technology. Storage devices based on carbon structures will be developed.

Storage Barriers

Hydrogen storage must meet a number of challenges before hydrogen can become an acceptable energy option for the consumer. The technology must be made transparent to the end-user—similar to today's experience with internal combustion gasoline-powered vehicles. Specific challenges include the following:

• **Current research and development efforts are insufficient**. A substantial research and development investment in hydrogen storage technologies will be required to achieve the performance and cost

Figure 6-4. Hydrogen Storage Buffer at SunLine Transit, California

targets for an acceptable storage solution. In particular, new media development is needed to provide reversible, low-temperature, high-density storage of hydrogen.

- **Low demand means high costs**. As there are few hydrogen-fueled vehicles on the road today, the more mature compressed and liquid hydrogen storage technologies are quite expensive. The initially low rates at which hydrogen end-use applications (particularly vehicles) will be introduced will present a challenge to the commercialization and cost reduction of hydrogen storage technologies.

Storage Strategies

Achieving the storage vision will require coordinated activities that address the challenges. In particular:

- **Develop a coordinated national program to advance hydrogen storage materials**. A fully funded national program is needed to improve the performance and reduce the costs of hydrogen storage. In particular, new materials that allow for chemical hydride and solid-state storage of hydrogen are needed. In addition, storage R&D must be elevated to a level that is commensurate with its importance. Because storage technologies are so integral to a successful hydrogen economy, R&D in this area must be increased to match R&D efforts in other segments of hydrogen development.

- **Develop a mass production process for hydrogen storage media**. Currently, no market force is driving efforts to reduce raw material costs and develop efficient mass production processes. Even the more mature compressed and liquid hydrogen storage technologies are expensive due to an absence of high-volume demand. Once hydrogen storage materials and technologies have been optimized in the lab, practical integrated storage systems must be developed and demonstrated.

The lack of low-cost and lightweight storage devices, as well as commercially available and cost-competitive fuel cells, interferes with the implementation of hydrogen as an energy carrier. For a hydrogen economy to evolve, consumers will need to have convenient access to hydrogen, and storage devices will be one of the keys. Better hydrogen storage systems will offer easy access to hydrogen for vehicles, distributed energy facilities, or central station power plants.

HYDROGEN CONVERSION

Conversion Status

Hydrogen can be used both in engines and in fuel cells. Engines can burn hydrogen in the same manner as gasoline or natural gas, while fuel cells use the chemical energy of hydrogen to produce electricity and thermal energy. Since electrochemical reactions are more efficient than combustion at generating energy, fuel cells are more efficient than internal combustion engines.

The use of hydrogen in engines is a fairly well developed technology, and new combustion applications are under development. Vehicles with hydrogen internal combustion engines are now in the demonstration phase, and the combustion of hydrogen blends is being tested. Fuel cells are in various stages of development. Current fuel cell efficiencies range from 40-50% at full power and 60% at quarter-power, with up to 80% efficiency reported for combined heat and power applications. A summary of hydrogen conversion technologies and applications are shown in Table 6-3.

Conversion Vision

Stakeholders identified the vision for hydrogen storage as follows:

Fuel cells will be a mature, cost-competitive technology in mass production. Advanced, hydrogen-powered energy generation devices such as combustion turbines and reciprocating engines will enjoy widespread commercial use. The commercial production, delivery, and storage of hydrogen will go hand in hand with the commercial conversion of hydrogen into valuable energy services and products, such as electricity and thermal or mechanical energy. The technologies for end-use will be well established. Current products embodying these technologies will provide safe, clean, and affordable energy services in all sectors of our global economy.

Conversion Barriers

All of today's conversion products, demonstration models, and prototypes possess some deficiencies; they cannot yet provide, at an affordable cost, the level and quality of energy services demanded by a broad base of consumers. While fuel cell technologies have generated much excitement, they are still in various stages of maturity. Most have

Table 6-3. Hydrogen Conversion Technologies and Applications

Technology	Application
Combustion	
Gas Turbines	• Distributed power • Combined heat and power • Central station power
Reciprocating Engines	• Vehicles • Distributed power • Combined heat and power
Fuel Cells	
Polymer Electrolyte Membrane (PEM)	• Vehicles • Distributed power • Combined heat and power • Portable power
Alkaline (AFC)	• Vehicles • Distributed power
Phosphoric Acid (PAFC)	• Distributed power • Combined heat and power
Molten Carbonate (MCFC)	• Distributed power • Combined heat and power
Solid Oxide (SOFC)	• Truck APVs • Distributed power • Combined heat and power

Figure 6-5. Hydrogen Fuel Cell Bus at SunLine Station, California

not been manufactured in large quantities and numerous performance issues—including durability, reliability, and cost—remain to be resolved. Combustion turbines and engines that use hydrogen or hydrogen/natural gas blends, already in use in both mobile and stationary applications, are much closer to satisfying these criteria than are fuel cells. Some of the barriers preventing the market development of hydrogen devices include:

- **No single fuel cell technology has met all the basic criteria for performance, durability, and cost**. Basic and applied research in materials science and electrochemistry is required to improve the design and operation of all fuel cell technologies and provide and ongoing basis for substantial cost reduction and performance improvements.

- **Fuel cells require enhanced materials, membranes, and catalysts to meet both engineering and cost criteria**. For all types of fuel cells except phosphoric acid fuel cells, reliability of performance and durability over extended hours of operation remain to be proven. Questions remain about the performance of all types of fuel cells under diverse climatic conditions and geographic locations. Manufacturing scale-up issues and the associated need to establish high-volume demand are major barriers in achieving cost reductions.

- **Research is needed to fill in critical knowledge gaps**. Researchers require better information about the flame characteristics of hydrogen combustion and the impacts of conversion technologies on reciprocating engine and turbine designs. Existing databases need to be populated with more performance data for hydrogen-burning engines and turbines operating over extended periods; performance data needs include efficiency, emissions, and safety, for both mobile and stationary applications.

- **Market and institutional barriers hinder development of cost-competitive hydrogen conversion devices**. Customers do not see a robust value proposition that convinces them to choose hydrogen conversion products. Substantial cost reduction will be essential—particularly without a bridging incentive or government mandate fostering use of hydrogen conversion products rather than lower-cost conventional fuels and products. In the absence of such policies, conventional fuels and conversion devices will continue to be the only practical option for consumers.

Conversion Strategies
Most of the paths forward involve research and development activities. In particular:

- **Continue research and development on fuel cell materials and engines**. Investing in efforts to increase fundamental understanding of current materials, interfaces, and processes will support important advances, such as improving and reducing the costs of the catalysts used in fuel cells, or developing materials that will improve thermal management and invite combined heat and power opportunities.

- **Enhance manufacturing capabilities for fuel cells**. Techniques are needed for handling high fuel cell production volumes and achieving better consistency and quality control. Advancements in this area are one of the surest means to achieving the large cost reductions needed to move fuel cells from niche to mass markets.

• **Collect more and better information on operating performance at existing demonstration sites**. Improved instrumentation and expanded data collection efforts are required to facilitate analysis of the full range of cost, efficiency, and emissions parameters for all mobile and stationary applications under a wider range of environmental conditions. At the same time, better market analysis is needed to provide the financial community with an improved understanding of the potential for fuel cells and hydrogen-using engines and gas turbines.

HYDROGEN APPLICATIONS

Application Status

Hydrogen can be used in conventional power generation technologies, such as automobile engines and power plant turbines, or in fuel cells, which are relatively cleaner and more efficient than conventional technologies. Fuel cells have broad application potential in both transportation and electrical power generation, including on-site generation for individual homes and office buildings.

Transportation Applications Today

Transportation applications for hydrogen include buses, trucks, passenger vehicles, and trains. Technologies are being developed to use hydrogen in both fuel cells and internal combustion engines. Nearly every major automaker has a hydrogen-fueled vehicle program, with various targets for demonstration between 2003 and 2006. The early fuel cell demonstration programs will consist of pilot plant "batch builds" of approximately 10 to 150 vehicles. Information obtained form these vehicle demonstrations will then be used to help determine how and when to advance to the next level of production.

Hydrogen-fueled internal-combustion engine vehicles are viewed by some as a near-term, lower-cost option that could assist in the development of hydrogen infrastructure and hydrogen storage technology. A key advantage of this option is that hydrogen-fueled internal-combustion engine vehicles can be made in larger numbers when demand warrants.

Stationary Power Generation Today
Stationary power applications include backup power units, grid management, power for remote locations, stand-alone power plants for towns and cities, distributed generation for buildings, and cogeneration (in which excess thermal energy from electricity generation is used for heat). Although commercial fuel cells are on the market, the industry is still in its infancy. Most existing fuel cell systems are being used in commercial settings and operate on reformate from natural gas. Widespread availability of hydrogen would allow the introduction of direct hydrogen units—simpler systems with lower cost and increased reliability.

In general, combustion—based processes, such as gas turbines and reciprocating engines, can be designed to use hydrogen either alone or mixed with natural gas. These technologies tend to have applications in the higher power ranges of stationary generation.

Portable Power Generation Today
Portable applications for fuel cells include consumer electronics, business machinery, and recreational devices. Many participants in the fuel cell industry are developing small-capacity units for a variety of portable and premium power applications ranging from 25-watt systems for portable electronics to 10-kilowatt systems for critical commercial and medical functions. Most of these portable applications will use methanol or hydrogen as the fuel. In addition to consumer applications, portable fuel cells may be well suited for use as auxiliary power units in military applications.

Application Vision
Stakeholders identified the vision for hydrogen applications as follows:
Hydrogen will be available for every end-use energy need in the economy, including transportation, power generation, industrial process heaters, and portable power systems. Hydrogen will be the dominant fuel for government and commercial vehicle fleets. It will be used in a large share of personal vehicles and light duty trucks. It will be combusted directly and mixed with natural gas in tur-

Figure 6-6. Various Sizes of Hydrogen Fuel Cells

bines and reciprocating engines to generate electricity and thermal energy for homes, offices, and factories. It will be used in fuel cells for both mobile and stationary applications. And it will be used in portable devices such as computers, mobile phones, internet hook-ups, and other electronic equipment.

Application Barriers

To achieve this vision for hydrogen applications, the following challenges will need to be overcome.

- **Transportation, stationary, and portable applications require technological and engineering solutions.** Transportation applications lack affordable and practical hydrogen storage with sufficient volumetric and gravimetric densities. The absence of a storage solution severely hinders investment in infrastructure development as different storage medium could result in substantially different infrastructure strategies. There is also a lack of reliable, inexpensive, and efficient reformation technologies.

- **Customers must accept hydrogen technologies and fuel cell vehicles.** Fuel cell vehicles are in the early stages of development and the first vehicles are likely to fall short of consumer expectations. By comparison, conventional internal-combustion engine vehicles have had the benefit of more than 100 years of technological refinement as well as relatively reliable, low-cost gasoline to power them.

- **Competition from other advanced technologies**. There are competing technologies, such as hybrid-electric vehicles and advanced conventional fuel vehicles that are cleaner than today's conventional vehicles. These advanced vehicles make hydrogen's case on environmental grounds weaker.

Application Strategies

Like other components of the hydrogen system, the major path forward focuses on advanced research and development. For transportation and stationary applications, developing low-cost and reusable fuel cell stacks and systems are required.

In addition, the following must be done:

- **Increase demonstrations significantly**. Demonstrations should showcase the near-term availability of multiple alternative technologies for distributed generation power parks.

- **Institute regulations, codes and standards to foster customer acceptance of the hydrogen vision**. Standard nationwide interconnection agreements are needed to enable connection to the current electrical grid without punitive costs, policies, or actions. Standard agreements and educational materials should be prepared for use by fire, insurance, and building code officials.

- **Develop public policies that encourage use of hydrogen as a fuel**. Convincing Americans to use hydrogen applications will require incentives such as cost-sharing demonstration, policies for price parity, and "rights of way" for hydrogen infrastructure (similar to those in the natural gas industry). The federal government should adopt national interconnection standards and ensure that distributed generation options are valued for their ability to utilize waste heat and achieve high efficiencies. Government should also provide incentives for investing in new technologies, such as tax credits for transportation, stationary, and portable hydrogen systems, and for hydrogen infrastructure development.

The ultimate aim is to enable consumers to use hydrogen energy devices for transportation, electric power generation in cities and homes, and portable power in electronic devices such as mobile phones and laptop computers. Once the cost and performance issues associated with hydrogen energy systems have been addressed, the next challenges will involve customer awareness and acceptance. Safety, convenience, affordability, and environmental friendliness are key consumer demands. Industry should focus its efforts on understanding consumer preferences and building them into hydrogen system designs and operations. Government (federal and state) should identify opportunities to use hydrogen systems in facilities for distributed generation, combined heat and power, and vehicle fleets.

CONCLUSIONS

One of the major issues that has to be addressed is ensuring that all of these components work together in a systems approach. For example, storage devices need to be compatible with conversion devices; production methods must be compatible with delivery systems; and applications must be compatible with consumer demands. In addition, industry and government need to work together to conduct public outreach, address safety concerns, and establish networks by which people can share information and outreach.

This roadmap defines a common set of objectives and activities agreed upon by government, industry, universities, national labs, environmental organizations, and other interested parties. Focusing resources on this common agenda will facilitate evaluation of a hydrogen economy and potentially stimulate investment in the development of a hydrogen energy system.

Development of hydrogen energy technologies represents a potential long-term energy solution with enormous benefits for America. A coordinated and focused effort is necessary to bring public and private resources to bear on evaluating the costs and benefits of the transition to a hydrogen economy. The immediate next step that emerges from this roadmap is the development of detailed research and development plans for each technology area.

[**Editor's Note**: This chapter represents an abridged and edited version of the U.S. Department of Energy report entitled *National Hydrogen Energy Roadmap* published by the U.S. DOE, November 2002. The U.S. DOE coordinator for the roadmap was Frank (Tex) Wilkins, Office of Hydrogen, Fuel Cells, and Infrastructure Technologies. The roadmap was developed by dedicated representatives of the hydrogen industry, in conjunction with personnel from the public sector and academia. The editor is grateful to the U.S. Department of Energy for permission to publish an abridged version of this report.]

Reference
[1]Proceedings from the vision meeting and other supporting material can be downloaded from www.eren.doe.gov/hydrogen.

Chapter 7
Biomass Technology Roadmap

A Glimpse of the Future

By 2020...

Electric utilities will use biomass to generate power at four times current levels, providing 5% of all energy use in the industrial and utility sectors.

Ten percent of the transportation fuels used in the country will be derived from biomass. People will refuel their vehicles with fuels such as biodiesel and ethanol developed from biomass like soybean oil, corn oil, switch grass, and other woody plants.

Almost 20% of the chemical commodities produced in the U.S. will be from biobased products—no more will companies be reliant on petroleum resources to produce certain chemical commodities.

Most landfills will be "tapped" and "mined" for natural gas and this gas will be used to generate heat and power for homes, schools, and industry.

New technologies will allow efficient production of biofuels, with less reliance on agricultural commodities and more reliance on grasses and woody plants.

INTRODUCTION

The purpose of this chapter is to outline a technology roadmap for the biomass industry. The roadmap focuses on research and development (R&D) activities and public policy measures for developing biobased fuels, power and products. This roadmap was produced through the Biomass Research and Development Technical Advisory Committee—a group representing a wide range of experts from the biomass industry.

The roadmap first identifies a vision for the biofuels, biopower, and biobased products industries. The roadmap then identifies the barriers preventing the industry from achieving this vision. Finally, the roadmap suggests technical and policy options that should be pursued to overcome these barriers.

THE ROADMAP PROCESS

The Biomass Vision

Environmentally sound biobased fuels, power, and products can make important contributions to U.S. energy security, rural economic development, and environmental quality. Understanding the benefits of these contributions, the Biomass Research and Development Technical Advisory Committee established challenging yet feasible goals for increasing the role of biomass technologies in the U.S. economy. These goals, shown in the text box below, project continued growth in biomass consumption from electricity production, and a significant increase in biomass consumption for the production of transportation fuel and biobased products. This roadmap presents recommended strategies, directions, and plans that should be used to achieve these goals.

Vision for a Biomass Future

- *Biopower*—Biomass consumption in the industrial sector will increase at an annual rate of 2% through 2030, increasing from 2.7 quads in 2001 to 3.2 quads in 2010, 3.9 quads in 2020 and 4.8 quads in 2030. Moreover, biomass use in electric utilities will double every ten years through 2030. Biopower will meet 4% of total industrial and electric generator energy demand in 2010 and 5% in 2020.
- *Biobased Transportation Fuels*—Transportation fuels from biomass will increase significantly from 0.5% of U.S. transportation fuel consumption in 2001 to 4% of transportation fuel consumption in 2010, 10% in 2020, and 20% in 2030.
- *Biobased Products*—Production of chemicals and materials from biobased products will increase substantially from approximately 12.5 billion pounds, or 5% of the current production of target U.S. chemical commodities in 2001, to 12% in 2010, 18% in 2020, and 25% in 2030.

The Biomass Roadmap

The roadmap was developed through a series of public meetings of the Biomass Research and Development Technical Advisory Committee. This committee was established by the Biomass R&D Act of 2000 (P.L. 106-224). The Committee developed this roadmap at the request of the U.S. Department of Energy and the U.S. Department of Agriculture as a tool to assist in biomass-related research planning and program evaluation. The focus of the public meetings was to identify the R&D and public policy activities that would be needed to achieve the biomass vision shown above.

This roadmap is organized by the major categories of R&D needed to achieve the biomass vision. These categories include:

- Feedstock Production
- Processing and Conversion
- Product Uses and Distribution

A section on public policy measures is also included. These measures should be adopted or evaluated to realize the full economic and environmental promise of biomass resources and technologies.

The majority of R&D identified in this roadmap will have crosscutting applications for biobased fuels, power and products. In other cases, R&D will be specific to one or more biobased applications. If implemented, the combination of R&D and public policy strategies outlined below will:

- Increase the scientific understanding of biomass resources and better tailor those resources for a variety of end-use applications;

- Improve sustainable systems for developing, harvesting, and processing biomass resources;

- Improve efficiency and performance in conversion and distribution processes and technologies for a host of products;

- Create the regulatory and market environment necessary for increased development and use of biobased fuels, power, and products;

- Improve environmental quality;

- Enhance access to biomass sources; and,

- Ensure U.S. world leadership in the development of biomass conversion and enabling technologies.

BIOMASS FEEDSTOCK PRODUCTION

Advances in feedstock production are extremely important. These advances have the potential for reducing the final cost of biobased fuels, power, and products. In addition, advances will allow the production of plants, trees, and residues with characteristics increasingly well suited for feedstocks. For example, genetically engineered feedstocks may allow higher yields of usable biomass per acre for fuel and other uses. New methods in erosion control, fertilization, and pre-processing can result in improved life cycle performance, sustainable practices, and enhanced feedstock production.

Feedstock production faces a number of challenges that currently

Figure 7-1. Alcohol Production at NREL from Non-Food Crops

hinder the ability of biomass communities to achieve vision goals. The main challenges are:

1. A better understanding of plant biochemistry and enzymes is needed;

2. Scientific methods to product and prepare plants and residues so that they meet specifications for end-use applications are needed; and,

3. Agronomic practices must be improved to increase efficiency and reduce the cost of biomass feedstock production and delivery, and to ensure crop sustainability.

Achieving the Biomass vision goals will require a change in the entire biomass production system including new and better methods for crop growth and management, harvesting, densification, transportation, storage, and pre-processing. It will require both small, more localized processing plants and/or larger scale ones that take advantage of economies of scale.

Advances in R&D can help to improve storage methods, expand the growth of crops for energy and other products, and assure the quality of feedstocks. At the same time, research into the agronomic, economic, and environmental impacts of harvesting lignocellulosic material must be established to ensure that these materials have beneficial life-cycle impacts.

Basic research can help to address broad needs for improving development of a number of feedstocks that can ultimately be used as resources for a wide range of biobased applications. At the same time, however, a better definition of high priority applications/products is needed to help growers and the research community focus more applied research in feedstock production.

Finally, continued advances in feedstock production research will face the special challenge of public acceptance. Improved methods for verifying the safety as well as the society and environmental benefits of genetically engineered plants are needed. Resulting data should be used to engage in broad, multi-party stakeholder dialogue to determine whether to and/or how best to commercialize these technologies. Regarding genetically engineered plants, coordinated leadership from key federal providers of R&D, such as the U.S. Department of Energy and the U.S. Department of Agriculture, will be critical to making these scientific advances, validating their success and safety, and improving public acceptance through dialogue and stakeholder engagement.

Achieving vision goals for biomass technologies will require significant advances in feedstock production. Ultimately, it will require an increased scientific understanding of methods for high yield, low input targeted crops produced in a sustainable and environmentally sound manner. The specific R&D objectives for feedstock production research are shown in Table 7-1. This table also identifies the Key Outcomes of the research and the potential impact in each of the three areas covered in this roadmap (Biofuels, Biopower, and Bioproducts).

Feedstock research should focus on a representative cross-section of biomass resources including: corn stover, dry land crops such as oilseeds, dedicated energy crops, and plant, animal, and other organic waste-based residues. Target crops should include oil and cellulose-producing crops that can provide optimal energy content and usable plant components. Key outcomes of advanced feedstock production research should produce several important results for the biomass communities. Examples include:

- Increased yield per acre
- Lower cost per ton of feedstock at plant gate
- Increased value for the outputs of biomass feedstocks
- Reasonable profit for growers, and
- Environmentally sound production of biomass.

Specific research objectives are presented below for items found in Table 7-1.

A.   Biotechnology and Plant Physiology
Objective One

Improve the technical understanding of plant biochemistry and enzymes and develop the ability to engineer enzymes within desired crops.

Ultimately, the ability to produce high value, environmentally sound biobased fuels, power and products may require lower cost feedstocks (i.e. crops, agricultural plant and animal residues). This feedstock production system will require profit for the grower. Examples of desir-

Table 7-1. Crosscutting Impacts of Feedstock Production R&D

Major R&D Needs	BioFuels Impact	Biopower Impact	Bioproducts Impact
Biotechnology, genetics, and plant physiology	H	H	H
• Improve basic science in plant genetics and biochemistry	M	M	M
• Improve chemical and biological processes for improved feedstocks	H	H	H
Optimize agronomic practices, including addressing land use availability and soil sustainability issues.	H	H	H
Optimize logistics for collecting, storing, and combining multiple feedstocks with diverse applications.	H	H	L
Key Outcomes			
• Identify high opportunity plant and residue feedstocks.	H	H	H
• Increase yield per acre	H	H	H
• Reduce feedstock cost per ton at plant gate	H	H	H
• Increase dollar value of biobased outputs	H	H	H
• Reasonable profit for growers	H	H	H
• Environmentally sound production of biomass	H	H	H

Notes: H = High Impact; M = Medium Impact; L = Low Impact.

able characteristics include high-energy content, increased yield, fast growth, and the ability to withstand drought or other stresses.

Increased knowledge of the metabolic pathways that lead to lignins, proteins, and other plant components is needed. Currently, however, there is a lack of understanding of plant biochemistry as well as inadequate genomic and metabolic data on many potential crops. Specifically, research to produce enhanced enzymes and chemical catalysts could advance biotechnology capabilities. Enzymes and catalysts are necessary to efficiently and cost effectively turn biomass feedstocks into biobased products, fuels, and power.

Research is needed to produce crops with desirable traits for both edible and industrial uses. For example, simply increasing the plant's natural production of a specific component already found in the crop, such as a protein or fatty acid, can increase the ultimate yield of that component and thereby improve efficiency and reduce the cost of bioenergy and biobased product processing and conversion.

Objective Two

Develop the chemical and chemical/biological pathways necessary to improve the energy density and chemical characteristics of delivered feedstocks.

The most valuable way to improve the availability and cost competitiveness of biobased fuels, power, and products is to develop advanced methods for overcoming the resistance of agricultural, forest-based, and urban feedstocks to enzymatic and fermentation treatments.

Current technologies for creating a treatable/fermentable product from available, environmentally appropriate biomass resources do not meet the economic needs of the industry. Examples of research needs include:

- Fundamental Structure of Lignocellulosic Materials—To improve growth rates, research is needed on the fundamental structure of lignocellulosic materials, including the chemistry of its cell wall structures, transport properties, and genetic potential. This research can provide the basis for developing a sufficient quantity of cost-competitive biomass feedstocks necessary to achieve vision goals for biobased fuels, power, and products.

- Cost-effective Pre-delivery Treatment Processes—Research should also include development and testing of cost-effective pre-conversion treatment processes to increase energy- and chemical-density of raw materials at the point of harvest.

B. Agronomic Practices
Objective Three

Optimize agronomic practices for sustainable biomass feedstock production.

Achieving vision goals will require research to improve and advance agronomic practices. Energy crops will compete for land with existing land uses such as traditional agriculture and forestry. Energy crops also offer the potential to expand the resource base to marginal agricultural land. Research must evaluate methods to ensure the availability of land for producing the biomass feedstocks necessary to achieve vision goals.

C. Feedstock Handling
Objective Four

Optimize logistics for collecting, storing and combining multiple feedstocks that can be applied for diverse applications in an environmentally sound manner.

There are a number of opportunities to improve the mechanical systems associated with feedstock handling so that biomass resources can be used for a wider variety of applications. Improvements in feedstock analysis and preparation technologies as well as mechanical harvesting and storage practices should help lower the cost of production and delivery of biomass feedstocks. Research is needed to advance existing technologies and processes in these areas as well as to develop new technologies. This research should enable the handling and storage of unique combinations of biomass feedstocks that are tailored for specific applications, without sacrificing the integrity of the feedstock. Finally, outreach and education are needed to improve public understanding of different applications for animal residue, and the relative environmental impacts of each. Specific examples of research needs include:

- Feedstock Density—Research is needed to improve/develop mechanical technologies that will increase both energy and physical density and reduce moisture content of plant and animal residue-based biomass feedstocks.

- Sensors—Quick, cost-effective systems for on-line real-time analysis and maintenance of plant and animal residue-based feedstocks must be developed. These systems should monitor and maintain feedstock quality through the collection, storage, and transportation phases of the product life cycle. They should provide producers with a better understanding of the quality of the feedstock. Additionally, systems should be developed to monitor growth so that harvesting can occur at the optimum time for processing and conversion.

- Best Practices for Harvesting and Storage—The biomass/agricultural communities must identify, develop, test, and implement best practices for cost-effective and environmentally sound pre-treatment, collection, storage, and transport of plant and animal residue-based biomass feedstocks. This should lead to improved plant and animal residue recovery, more effective separation, improved handling and storage technologies/procedures, and reduced environmental impacts.

PROCESSING AND CONVERSION

Technical advances in biomass processing and conversion technologies will improve conversion efficiencies and increase the output of useful energy and product per unit of input while reducing negative environmental impacts. However, research is needed to facilitate commercially viable and environmentally sound biomass processing and conversion systems. As shown in Table 7-2, several of these research areas crosscut biobased fuels, power and products whereas others are specific to one or more end-uses.

Expanding the use of biomass for non-food and feed purposes will benefit farmers and rural areas only indirectly and modestly. A more significant development would occur if farmers were able to produce the biofuels or bioproducts themselves, either on the farm or as owners in a local production plant.

Industry, universities and the national laboratories should work together on pilot plant facilities that focus on evaluating and developing processing technologies for bioenergy and biobased products using a variety of raw material resources. Existing pilot plant facilities should be

Table 7-2. Crosscutting Benefits of Processing and Conversion R&D

Major R&D Needs Impact	Biofuels Impact	Biopower Impact	Bioproducts
Thermochemical Conversion			
• Co-firing	L	H	L
• Direct combustion	L	H	L
• Gasification	L	H	L
• Anaerobic Fermentation	L	H	L
• Modular Systems	L	H	L
• Pyrolisis	L	H	M
Bioconversion			
• Physical/Chemical Pretreatment	H	M	H
• Fractionation and Separation	H	M	H
• Residual Solids and Liquids	H	M	H
• Chemical/Enzymatic Conversion	H	M	H
• Catalytic and Chemical Conversion	H	M	H
• Inhibitory Substances	H	M	H
• Separation and Purification	H	M	H
• Biomass Fermentation and Hydrolysis	H	M	H
• Syngas Fermentation	H	M	H
Biorefinery Integration	H	H	H

Notes: H = High Impact; M = Medium Impact; L = Low Impact. Readers should refer to the roadmap report for detailed descriptions for each research area.

Figure 7-2. Biomass Gasifier Using Sugarcane Residue

inventoried and used to prove and optimize production techniques and economics. Special consideration should be given to agriculture and forestry-based cooperatives in licensing technologies development with government support. Efforts should be made to identify existing facilities that can be converted into or enhanced as biorefineries. Finally, an emphasis should be put on rural-based biorefineries. Each of the items in Table 7-2 is presented below in more detail.

A.   Thermochemical Conversion Pathways
Objective One

Develop cost-effective, environmentally sound thermochemical conversion technologies to convert biomass feedstocks into useful electric power, heat and potential fuels and products.

Biomass resources are currently used to produce electric power and/or heat at some industrial facilities across the United States. In addition, biomass is a minor resource for electric utilities across the country. The ability to overcome several significant barriers in thermochemical conversion, however, could increase the role that biomass systems play in providing heat, power, fuels, and products. The initial thrust should be the consumption of residue biomass. Examples of needs in thermochemical conversion research include:

- Improvements in biomass gasification technologies to enable the conversion of a wide range of feedstocks, starting with residue biomass

- The integration of conversion systems with power generation equipment, • expansion of capabilities to convert low quality gas into electricity

- Methods to overcome technical barriers to thermochemical conversion, such as tar removal, prior to firing biogas in a turbine system, and

- Analytical studies on costs, performance, and life-cycle emissions as well as scalability analysis (lab to commercial scale) of thermochemical conversion processes.

Some more detailed areas of research are:

Co-firing
The environmental benefits of co-firing, as demonstrated in some industrial facilities, include fossil fuel replacement and reduction of negative environmental impacts. In addition, co-firing provides near-term demand for biomass feedstocks that will help develop the infrastructure needed to produce and deliver these resources to stand-alone biomass electric generation facilities and integrated biorefineries. Previous research has led to demonstration projects which show co-firing to be a technically viable option for utilities at the current time. Although technically viable, its use is not widespread. Opportunities still remain to improve operating efficiencies. Greater research as well as increased use in industry could help lead to these improvements. Increased technology transfer and demonstrations from forest products applications could help to increase the spread of co-firing technologies.

Direct Combustion

Direct combustion of biomass is currently in use in the United States. Improvements could still be made, however, to improve operating efficiencies.

Biomass Gasification

Biomass gasification technologies are currently in place. However, there remain a number of technical and economic hurdles to improve their cost competitiveness with other technologies. Specifically, research is needed to reduce the capital costs and improve the operating efficiencies of gasification systems. In addition, research should be performed to enable gasification of a wider range of resources, such as forest and agricultural residues as well as to expand the development and application of black liquor gasification. Gasification technologies should be designed for integration with generating turbines and biorefineries.

Anaerobic Fermentation Gases

Power and fuels can currently be produced from anaerobically generated gases. These include landfill gases, anaerobic digestion of animal manure and food/feed/grain products and by-products, use of wastewater treatment digestion gas, sludge and sewage treatment gases, and other sources. Methane emitted from biomass waste is a potent greenhouse gas, with a global warming potential 21 times that of Carbon dioxide. Over 600 million tons of carbon equivalent methane are produced annually in the United States. There remain opportunities, however, for greater application of anaerobic fermentation. Research should address needs including increasing the rate of decay for residues used in anaerobic fermentation systems. Research is also needed to reduce capital costs and improve operating efficiencies of these systems. Moreover, low intensity methane should be viewed as a resource instead of a waste product. Systems for the use of methane from 10—300 Btu/cu. ft. are technically feasible and should be developed and demonstrated. Research on integrating systems with anaerobic digestion provides another opportunity for synergies between technologies.

Modular Systems

Modular systems are currently available and used in the United States and internationally. However, advances need to occur in the development of modular systems and distributed small-scale generation of less than 1 MW. Research to make these systems cost competitive is needed including research to reduce their capital costs. Systems should be developed that can consume small quantities of organic waste or dedicated resources for distributed generation of power and heat locally for use on-farm, on-site, and in small industrial systems. The alternatives developed could include integration of modular biomass systems with fuel cells, microturbines, and other distributed systems. Resources include food/feed/grain processing plant residue, fats and oils, nutshells, corncobs, tomatoes, carrots, fruit, rice hulls, as well as uncontaminated urban wood residue and farm animal waste. Significant R&D opportunities in this area are the development of scaled-down, skid-mounted or mobile installations and fuel concentrators to increase energy density. Significant opportunities for modular systems exist in low value by-products from grain, soy, wood and other processing systems, and in farm and forest residues where the high cost of transporting biomass to larger facilities can be avoided. Rural communities and farmers could benefit if modular systems are developed that can be deployed to offset

Figure 7-3. Biomass to Electricity Facility

power costs in grid-based systems in the United States. Industry standards for grid connection should be simplified and new standards developed so that modular biomass systems can be easily connected to the grid. The waste biomass from any biorefinery that has no other value will still be able to be converted into electricity.

B.   Bioconversion
Objective Two

Develop economically viable and environmentally sound bioconversion processes/technologies for commercial application of a range of biobased fuels and products.

Advances in biochemical conversion processes will increase the variety of biofuels and biobased products that can be cost-competitively produced from biomass resources. For example, research is needed to enable conversion of multiple sugar streams and lignocellulosic materials to useful fuel or value-added products. In addition, research is needed to develop enzymatic pre-treatment methods for increasing the efficiency of biofuels production.

Examples of bioconversion research needs are in two general categories: Processing and Conversion. Improved methods and technologies for processing biomass feedstocks are needed to increase both the economics and technical capabilities of bioconversion systems. Specific research needs include:

Physical and Chemical Pretreatment Prior to Fermentation

Improvements are needed to improve physical and chemical pretreatment of biomass feedstocks prior to fermentation. This may include new enzymes and/or new methods for enzyme pretreatment.

Biomass Fractionation and Separation Technologies

Traditional agriculture and forest crops, urban waste, and crop residues represent a major source of readily available complex proteins, oils, and fatty acids as well as simple and complex sugars to be used as raw materials. These materials are available at low cost in localities across the United States. There is a need for R&D to develop low-cost chemical and biological processes including new chemistry and thermochemical synthesis that can break down these molecules and separate the resulting components into purified chemical streams. New concepts need to be developed and past separation technologies should be reexamined to reduce costs in downstream processes.

Utilization of Residual Solids and Liquids

Currently, residual biomass resources exist in the form of plant, animal, and other residues. These residues can be used to develop value-added fuels, chemicals, materials, and other products. Research should be performed to develop cost effective methods for processing solid and liquid residues such that they become economically viable biomass resources.

Chemical/Enzymatic Conversion Processes

New cost-effective methods of chemical/enzymatic conversion should be developed and tested to make greater utilization of biomass resources. Specifically, scalability analysis of biochemical conversion systems is needed.

Catalytic and Chemical Conversion

Catalytic and chemical methods for converting vegetable oils and animal fats into biodiesel are currently in use. R&D is necessary to improve the efficiency of these processes, to develop new processes, and to make processes more cost-competitive with non-biobased products.

Inhibitory Substances in Sugar Streams

Research is needed to overcome the barriers associated with inhibitory substances in sugar streams. For example, methods to enable removal of catalytic inhibitors should be developed. Similarly, catalysts could be engineered to enhance their tolerance.

Separations and Purification

Research is needed on engineering and biological principles as well as combinations of both to improve feedstock separation and product purification.

Biomass Fermentation and Hydrolysis

Research is needed to enhance the fermentation and hydrolysis of fiber, oil, starch, and protein fractions of crop components and processing by-products. In addition to the need to enable more rapid conversion of cellulose to a fermentable substrate, there is a need to develop new fermentation technologies to enable production of base chemicals and chemical intermediates from the wide range of existing crop components.

Syngas Fermentation

Research to improve catalytic synthesis of gases to chemicals as well as to improve pyrolysis to produce chemicals is also needed. Processing systems must optimize both mass transfer of oxygen and nutrients for bioorganisms and the fermentor environment.

C.   Biorefinery Integration

Objective Three

Advance the development of biorefineries that 1) efficiently separate biomass raw materials into individual components, and 2) convert these components into marketable products, including biofuels, biopower, and conventional new bioproducts.

Biorefineries already exist in some agricultural and forest products facilities (e.g. corn wet milling and pulp mills). These systems can be improved through better utilization of residues; new biorefineries can be enhanced by applying the lessons learned from existing facilities to comparable situations.

Biorefineries can become markets for locally produced biomass resources and simultaneously provide a local and secure source of fuels, power, and products. Optimized systems like biorefineries will potentially use complex processing strategies to efficiently produce a diverse and flexible mix of conventional products, fuels, electricity, heat, chemicals, and material products from biomass. Examples of research needs include but are not limited to:

- Further evaluation, development and deployment of the biorefinery concept for local and regional markets.

- Utilization of existing biomass processing and conversion facilities in the development of biorefineries.

- Development of new cost-competitive biomass technology platforms for additional biorefinery concepts.

- Bioconversion of sugars to products such as polyols or other products that can be used to produce chemicals, materials, or other biobased products.

- Development and commercialization of the conversion of vegetable oils to hydraulic fluids, lubricants, and monomers for a wide variety of uses in plastics, coatings, fibers, and foams to enable a biodiesel / bioproducts biorefinery.

- Development of alternatives to petroleum-based chemicals, polymers, plastics, and synthetic fibers.

- Development of alternatives to petroleum-based additives in the polymer industry including dyes, stabilizers, and catalysts.

- Creation of rural-based biorefineries that are modular, and produce high value products; residual waste from the biorefinery should be converted into electricity and useful heat.

In addition, efforts should be made to perform biorefinery pilot plant demonstration projects. Industry, universities, and national laboratories should work together on pilot plant facilities that focus on evaluating and developing processing technologies for bioenergy and biobased products using a variety of raw material resources. Existing pilot plant facilities should be inventoried and used to prove and optimize production techniques and economics. Special consideration should be given to agriculture and forestry-based cooperatives in licensing technologies developed with government support. Efforts should be made to identify existing facilities that can be converted into or enhanced as biorefineries.

Product Use and Distribution

The Biomass R&D Act of 2000 encourages the development of environmentally sound biobased fuels, chemicals, building materials, electric power, or heat. However, there are a number of barriers to the development, distribution and application of these technologies that will require R&D solutions. Moreover, simply increasing the use of biomass to produce electric power, heat, or other useful products is not the ultimate goal. Ultimately, research should enable a higher level of output of useful fuels, power, and products per ton of biomass inputs. This section describes research strategies to increase the use, efficiency, and sustainable of environmental sound biobased fuels, power and products in the economy.

There are three major objectives of this research:

1. Advance the understanding of biomass applications to expand existing markets, create new markets, and improve product distribution for environmentally sound bioenergy and biobased products.

2.   Identify and develop high value products from biomass feedstocks.

3.   Identify and develop distribution systems, and locate processing and conversion facilities in proximity to biomass resources, to maximize rural development and minimize negative environmental impacts.

Before the opportunities available from biomass technologies can be fully realized, targeted research activities are needed to improve markets and distribution systems for environmentally sound bioenergy and biobased products. Specific research needs include:

- *Biofuels Utilization Research*—Research must examine the fundamental properties of biofuels in pure form and in combination with petroleum-based fuels. For example, in the case of ethanol, fundamental research could help overcome questions of vapor pressure, ozone impacts, ethanol life-cycle impacts, and transportation.

- *Properties of Biofuels*—Research and testing activities are needed in several areas to improve the properties and marketability of biofuels. These include:
  — reducing the volatility of ethanol
  — increasing the flash point of biodiesel, and
  — improving the gel/pour point of biodiesel and reducing NO_x emissions

- *Ethanol Distribution in Pipelines*—Tests and demonstration projects on transporting ethanol by pipeline are needed. These activities must emphasize overcoming the problem of vapor pressure as well as decrease NO_x and ozone emissions.

- *Biorefinery Pilot Plant Demonstration Projects*—Industry, universities, and national laboratories should work together on pilot plant facilities that focus on evaluating and developing processing technologies for bioenergy and biobased products using a variety of raw material resources. Existing pilot plant facilities should be inventoried and used to prove and optimize production techniques and economics. Special consideration should be given to agriculture and forestry-based cooperatives in licensing technologies developed with government support. Efforts should be made to identify existing facilities that can be converted into or enhanced biorefineries. Finally, and emphasis should be put on rural-based biorefineries.

- *Gasification*—Innovative technologies need to be developed to solve the problems of animal litter and manure.

- *Hydrogen*—There are opportunities for biomass resources and technologies to contribute to a future hydrogen economy. Advances are needed in gasification, purolysis, and fermentation technologies to produce hydrogen form biomass corps, plant residues, or animal wastes. Federal programs involved in biomass research should coordinate closely with hydrogen research programs to identify and develop opportunities for using biomass to produce hydrogen. Research related to the use and distribution for hydrogen should be taken on by the federal hydrogen program.

- *Standards for Biobased Products*—Consumers should be able to make informed decisions regarding the performance, biodegradability, and other characteristics of biobased products versus competing products. Research should be performed to develop standards for biodegradability of biobased products. Performance standards should be established for biodegradability that are superior to existing standards for competing fossil-based products.

PUBLIC POLICY MEASURES

A number of public policy measures can be implemented to improve the status of biomass technologies in the marketplace. For example, consistent long-term federal policies are necessary to ensure the availability of loans and investment funding, encourage venture capital investment, and provide a sound footing for the development of new technologies. Current incentives, such as the ethanol tax incentive, have catalyzed the development of the fuels industry. To maintain the growth of the industry, equitable financial incentives, which would include tradable tax credits, should continue and incentives for other bioenergy and biobased products should be investigated. For example, definitions of

biomass used in the U.S. Tax Code (e.g., Section 45) should be broadened
to ensure that the range of environmentally sound biomass resources are
included; non-sustainable resources should be excluded. Incentives
should apply to both existing and new technologies and facilities.

Moreover, increased integration and/or coordination is needed
between the U.S. DOE and the USDA, and other federal agencies in
performing bioenergy and biobased products research, working with
industry to identify research priorities, and transferring research results
to industry. Both the EPA and the U.S. Department of the Interior, as well
as states and counties, should be involved in ensuring the greatest positive results for the environmental and the use of public lands.

The following specific policy proposals were identified as having
potential to advance environmentally sound biomass technologies and
achieve the goals set forth in the vision. The executive branch of the
government should assess the economic and environmental benefit that
these policies could produce and clearly recommend to the legislative
branch if and how to proceed with these issues. The following are general objectives for these policies:

1.   Promote the commercialization of successfully demonstrated envi-
     ronmentally sound biobased technologies.

2.   Outline the institutional policy changes needed to remove the bar-
     riers to economically sound development of sustainable biomass
     systems.

3.   Ensure that the biomass technologies developed are environmen-
     tally sound and move the country in the direction of a sustainable
     biomass system.

4.   Enhance opportunities for rural economic development.

To achieve these objectives, eight (8) specific areas are presented:

1.   *Economic Analysis.* Currently, there are many agricultural and en-
     ergy related policies, incentives, and other programs that may or
     may not encourage greater use of biomass resources and technolo-
     gies. An economic analysis of policies to promote the commercial-
     ization of successfully demonstrated environmentally-sound

biobased technologies should be prepared and widely distributed. This will help to focus future activity in sound policy and incentive development by providing a basis for knowledgeable choices among alternatives and allowing the cost of difficult to quantify benefits, such as energy security and environmental benefits, to be understood.

2. *Life-Cycle Assessment.* Life cycle cost, resource use, and environmental impact assessments must be performed for specific energy resources, chemicals and other products produced from biomass. The results should be compared to similar life cycle analyses on conventional processes for producing energy and chemicals to evaluate the relative costs and benefits of each. Life cycle stages should begin with resource production and continue through transportation, processing, conversion, end-use, and disposal/recycling. This will provide a balanced and meaningful comparison between biobased processes and competing processes in terms of both internal and external costs and benefits. The results of life cycle analyses should also be used to identify where costs and negative environmental impacts can be reduced and, subsequently, to test methods for reducing those costs. Moreover, life cycle cost and benefits of biomass technologies should be a component of public education. Greenhouse gas emission offsets should be considered when conducting economic and life cycle assessment.

3. *Procurement and Markets.*

   a. <u>Federal procurement</u>—The use of biobased fuels, power, and products should be encouraged through procurement standards for federal fleets, renewable energy purchasing requirements for federal facilities, and procurement requirements for biobased products.

   b. <u>Performance standards</u>—Performance standards should be developed for bioenergy and bioproducts with ASTM and other standards organizations.

   c. <u>Renewable portfolio standard</u>—The use of biopower should be encouraged through a renewable energy standard that applies to all electricity sellers at retail.

d. Renewable fuels standards—The use of renewable fuels should be encourage through a renewable fuels standard that applies to all transportation fuels.

e. Biofuels in watercraft—Requirements should be put in place for the use of biolubricants and biofuels in all watercraft.

4. *Regulatory Measures.*

a. Review current authorities—Identify existing federal and state authorities that can be used to facilitate, or act as barriers to, early adoption of environmentally-sound biobased technologies and products. This should include a survey of relevant policy and regulatory issues across federal and state governments (e.g. agriculture, forestry, energy, air, water, soil, environmental, grid access, recycling standards, access to biomass sources). Priority should be given to processing biomass related permit applications at environmental agencies. The DOE and USDA should be charged with the task of coordinating an interagency working group on biobased technologies with the potential to positively impact the production of biobased power, fuels, and products. There should not be discrimination among existing and new technologies and facilities.

b. Develop regulatory certainty—While it is essential to periodically review and improve regulatory standards and programs to reflect current knowledge and market experience, regulatory uncertainty has kept biomass-related industries and financial institutions from investing in biomass technologies. To the extent possible, regulatory systems should be consistent over the long term.

c. Removal of barriers to distributed generation—DOE and USDA should work together to ensure that investment in environmentally sound biomass technologies is not impeded through the use of unduly burdensome interconnection standards, inflated insurance requirements, or other artificial market barriers. Equitable access to the electricity grid for small, geographically dispersed power generators should be ensured.

d. Methane gas—The government should identify and develop regulatory mechanisms to maximize the use of recovered methane from landfills, wastewater treatment facilities, confined are feedlot operations, and other sources for electricity generation.

e. EPA New Source Review—Evaluate impacts of EPA's New Source Review restrictions for cofiring with biomass and revise them to encourage environmentally sound biopower development.

5. *Incentives*

a. Federal incentives—The government should encourage the use of biobased fuels, power and products through the use of equitable financial incentives, which include tradable tax credits, investment credits, and depreciation schedules. Federal incentive programs should favor farmer-owned production facilities (e.g. the CCC match for expanded ethanol and biodiesel facilities). Similarly, USDA energy efficiency and renewable energy incentives under the Energy title of the Farm Bill should favor on-farm, or farmer-owned facilities. These energy efficiency improvements should include changes in cultivation and livestock practices that lower pollution as well as measures to treat pollutants at "the end of the pipe." Federal incentives should not subsidize the waste disposal costs of businesses. In addition, federal incentives for methane-to-electricity generation should be allotted on a per ton of manure disposed of basis rather than per kilowatt-hour generated.

b. Financial Support for Existing Facilities—While incentives are often designed to drive investment in new facilities, many existing biomass based facilities are underutilized and warrant financial support to achieve full commercialization.

c. Public Benefit Funds—The use of biopower could be encouraged through financial incentives collected through non-bypassable surcharge on all retail electricity sales to support energy efficiency and renewables. Such funds should be available to all electricity users connected to the electric grid.

6. *Biomass Resource Supply*
    a. <u>Inadequate rural infrastructure</u>—Transportation systems (road and rail) need to be improved to facilitate the cost-competitive transport of biomass feedstocks from the farm and forest to the point of conversion.

    b. <u>Enhancing the Supply of Biomass</u>—Some of the land set aside by the Conservation Reserve Program (CRP) and the reduction of fuel loads in overly dense and overstocked forests, could be considered as sources of biomass for use in biopower, biofuels, and biobased products. For example, some CRP lands may be suitable for harvesting perennial grasses, trees, and producing energy crops while preserving soil and providing other benefits including wildlife habitat, carbon storage, and clean water.

7. *Education and Outreach*
    a. <u>Technology Transfer</u>—Federal agencies should assist in demonstrating novel biomass technologies. Examples include integrated biorefineries that demonstrate the cost-effective sustainable production of a host of biobased fuels, power, and products.

    b. <u>Perceptions of Biomass</u>—Many in the public are not aware of the benefits and ease of using biomass technologies and biobased products. Education and promotional campaigns are needed to communicate the performance and reliability associated with environmentally sound biomass technologies. Science-based education and outreach programs and materials should be developed that are directed toward classroom teaching and consumer education. K-12 outreach programs should also be developed to promote the environmental benefits of biomass technologies and products, and to explain how they compare with existing resources.

    c. <u>Centers of Excellence</u>—Centers of Excellence should be established at colleges and universities to enhance the nation's biobased products and bioenergy research capacity. These Centers of Excellence must play an important role in the education and training of the engineers, scientists, and business leaders that will catalyze the formation of the biobased product and bioenergy sector of the U.S. and world economy. Finally, these Centers of Excellence should contribute to local and regional biobased products and bioenergy economic development initiatives.

d. Biobased Logo/Labeling—A broad group including biobased industry, non-governmental organizations, etc., should develop a biobased logo to be associated with all environmentally sound biobased products, fuels, and power.

8. *R&D Investment*

a. Technology verification and financial risk—Pre-commercialization costs are very high, making it difficult for small start-up companies to develop and produce new biobased products on a large scale. Government should invest in these technologies and provide the financing often unavailable due to lack of market experienced.

b. Fostering innovation—Networks should be established to provide technical and/or grant assistance to innovators in the biomass field.

c. Risk Sharing—Government should encourage non-federal involvement in R&D by requiring some non-federal economic participation in government funded university research and other R&D efforts.

CONCLUSIONS

This roadmap presents a technical look at the research and development needs identified by the biofuels industry for successfully accomplishing the biomass vision. The vision itself represents a significant step forward for development of biofuels and biobased product markets.

In addition to calling for more RD&D in the biomass sector, the roadmap identified a host of public policy initiatives that could foster market growth. The roadmap also calls for a more systematic consideration of our energy systems so that biomass can be accounted for as we look to alternative energy futures.

[**Editor's Note**: This chapter represents an abridged and edited version of the report entitled Roadmap for Biomass Technologies in the United States published by the Biomass R&D Technical Advisory Committee in December 2002. This committee is part of the Biomass Research and Development Initiative and administrated by the National Biomass Coordination Office (which is run out of the U.S. DOE and is staffed by Departments of Energy and Agriculture). The editor is grateful to the Committee for permission to publish an abridged version of this report.]

APPENDIX 7-1:
BIOMASS RESEARCH AND DEVELOPMENT
TECHNICAL ADVISORY COMMITTEE MEMBERS

Roger Beachy, Donald Danforth Science Center
Robert Boeding, National Corn Growers Association
Dale Bryk, Natural Resources Defense Council
Robert Dorsch, Dupont
Glenn English, Jr., National Rural Electric Cooperative Association
Thomas W. Ewing, Davis and Harman, LLP
Carolyn Fritz, Dow Chemical Company
Stephen Gatto, Bionol Corporation
Brian Griffin, Oklahoma Secretary of Environment
Pat Gruber, Cargill Dow LLC
William Guyker, Life Fellow—IEEE
John Hickman, Deere & Company
Walter Hill, Tuskegee University
William Horan, Horan Brothers Agricultural Enterprises
Jack Huttner, Genencor International, Inc.
F. Terry Jaffoni, Cargill, Inc.
Michael Ladisch, Purdue University
David Morris, Institute for Local Self Reliance
William Nicholson, Potlatch Corporation, Retired
Edan Prabhu, FlexEnergy
William Richards, Richards Farms, Inc.
Philip Shane, Illinois Corn Marketing Board
Larry Walker, Cornell University
John Wootten, Peabody Energy
Michael Yost, Yost Farm, Inc.
Holly Youngbear-Tibbetts, College of Menominee Nation

APPENDIX 7-2:
DEFINITIONS AND TERMS FROM
ORIGINAL BIOMASS ROADMAP DOCUMENT

Agronomy: The science of plant production and soil management.

Anaerobic: Life or biological processes that occur in the absence of oxygen.

Biobased Product: Commercial or industrial products, other than food or feed, derived from biomass feedstocks. Many of these products possess unique properties unmatched by petroleum-based products or can replace products and materials traditionally derived from petrochemicals.

Biocatalyst: Usually refers to enzymes and microbes, but it can include other catalysts that are living or that were extracted from living organisms, such as plant or animal tissue cultures, algae, fungi, or other whole organisms.

Biochemical Conversion Process: The use of living organisms or their products to convert organic material to fuels.

Biodiesel: Conventionally defined as a biofuel produced through transesterification, a process in which organically- derived oils are combined with alcohol (ethanol or methanol) in the presence of a catalyst to form ethyl or methyl ester. The biomass- derived ethyl or methyl esters can be blended with conventional diesel fuel or used as a neat fuel (100% biodiesel). Biodiesel can be made from soybean or rapeseed oils, animal fats, waste vegetable oils, or microalgae oils. Note: Biodiesel can in certain circumstances include ethanol-blended diesel. This is an evolving definition.

Bioenergy: Useful, renewable energy produced from organic matter. The conversion of the complex carbohydrates in organic matter to energy. Organic matter may either be used directly as a fuel processed into liquids and gases, or be a residual of processing and conversion.

Biofuels: Fuels made from biomass resources, or their processing and conversion derivatives. Biofuels include ethanol, biodiesel, and methanol.

Biogas: A methane-bearing gas from the digestion of biomass.

Biomass: Any organic matter that is available on a renewable or recurring basis, including agricultural crops and trees, wood and wood wastes and residues, plants (including aquatic plants), grasses, residues, fibers, animal wastes, and segregated municipal waste, but specifically excluding unsegregated wastes; painted, treated, or pressurized wood; wood contaminated with plastic or metals; and tires. Processing and conversion derivatives of organic matter are also biomass.

Biopower: The use of biomass feedstock to produce electric power or heat through direct combustion of the feedstock, through gasification and then combustion of the resultant gas, or through other thermal conversion processes. Power is generated with engines, turbines, fuel cells, or other equipment.

Biorefinery: A processing and conversion facility that (1) efficiently separates its biomass raw material into individual components and (2) converts these components into marketplace products, including biofuels, biopower, and conventional and new bioproducts.

Biotechnology: A set of biological techniques developed through basic research and now applied to research and product development. In particular, biotechnology refers to the use by industry of recombinant DNA, cell fusion, and new bioprocessing techniques.

British Thermal Unit: Measure of energy based on the amount of heat required to raise the temperature of one pound of water from 59°F to 60°F at one atmosphere pressure.

Cellulose: The main carbohydrate in living plants. Cellulose forms the skeletal structure of the plant cell wall.

Co-Firing: The simultaneous use of two or more different fuels in the same combustion chamber of a power plant.

Cogeneration: The sequential production of electricity and useful thermal energy from a common fuel source. Reject heat from industrial processes can be used to power an electric generator (bottoming cycle). Conversely, surplus heat from an electric generating plant can be used for industrial processes, or space and water heating purposes (topping cycle).

Combined Cycle: Two or more generation processes in series, configured to optimize the energy output of the system.

Commercial Sector: An energy-consuming sector that consists of service-providing facilities of businesses; federal, state, and local governments; and other private and public organizations, such as religious, social, or fraternal groups. The commercial sector includes institutional living quarters.

Conservation Reserve Program: A voluntary USDA program whereby agricultural landowners can receive annual rental payments and cost-share assistance to establish long-term, resource conserving covers on eligible farmland. The Commodity Credit Corporation (CCC) makes annual rental payments based on the agriculture rental value of the land, and it provides cost-share assistance for up to 50 percent of the participant's costs in establishing approved conservation practices. Participants enroll in CRP contracts for 10 to 15 years. The program is administered by the CCC through the Farm Service Agency (FSA), and program support is provided by Natural Resources Conservation Service, Cooperative State Research and Education Extension Service, state forestry agencies, and local Soil and Water Conservation Districts.

Corn Wet Milling: A wet mill ethanol plant steeps (soaks in warm water) the corn. This enables separation of the germ, oil, starch, etc.

Densification: A mechanical process to compress biomass (usually wood waste) into pellets, briquettes, cubes, or densified logs.

Electric Utility: A corporation, person, agency, authority, or other legal entity or instrumentality that owns and/or operates facilities for the generation, transmission, distribution, or sale of electric energy primarily for public use. Utilities provide electricity within a designated franchised service area and file form listed in the Code of Federal Regulations, Title 18, Part 141. Includes any entity involved in the generation, transmission, or distribution of power.

Energy Crops: Crops grown specifically for their fuel value. These crops may include food crops such as corn and sugarcane, and nonfood crops such as poplar trees and switchgrass.

Energy Density: The energy content of a material measured in energy per unit weight of volume.

Environmentally Sustainable: An ecosystem condition in which biodiversity, renewability, and resource productivity are maintained over time.

Enzyme: A protein that acts as a catalyst, speeding the rate at which a biochemical reaction proceeds but not altering the direction or nature of the reaction.

Ethanol: Ethyl alcohol produced by fermentation and distillation. An alcohol compound with the chemical formula CH_3CH_2OH formed during sugar fermentation.

Feedstock: Any material converted to another form or product.

Fermentation: The biological conversion of biomass.

Forest Residues: Material not harvested or removed from logging sites in commercial hardwood and softwood stands as well as material resulting from forest management operations such as pre-commercial thinnings and removal of dead and dying trees.

Fossil Fuel: Solid, liquid, or gaseous fuels formed in the ground after millions of years by chemical and physical changes in plant and animal residues under high temperature and pressure. Oil, natural gas, and coal are fossil fuels.

Fuel Cell: A device that converts the energy of a fuel directly to electricity and heat, without combustion.

Gasification: A chemical or heat process to convert a solid fuel to a gaseous form.

Genetics: The study of inheritance patterns of specific traits.

Genetically Engineered Organism: An organism developed by inserting genes from another species.

Genomics: The study of genes and their function.

Greenhouse Gases: Gases that trap the heat of the sun in the Earth's atmosphere, producing the greenhouse effect. The two major greenhouse gases are water vapor and carbon dioxide. Other greenhouse gases include methane, ozone, chlorofluorocarbons, and nitrous oxide.

Grid: A system for distributing electric power.

Grid Connection: Joining a plant that generates electric power to an electric system so that electricity can flow in both directions between the electric system and the plant.

Hydrolysis: Conversion of biomass into sugars and sugar substrates via chemical or biological processes or through biocatalysis.

Industrial Sector: An energy-consuming sector that consists of all facilities and equipment used for producing, processing, or assembling goods. The industrial sector encompasses manufacturing; agriculture, forestry, and fisheries; mining; and construction.

Inorganic Compounds: A compound that does not contain carbon chemically bound to hydrogen. Carbonates, bicarbonates, carbides, and carbon oxides are considered inorganic compounds, even though they contain carbon.

Kilowatt: (kW) A measure of electrical power equal to 1,000 Watts. 1 kW = 3,413 Btu/hr = 1.341 horsepower.

Kilowatt hour: (kWh) A measure of energy equivalent to the expenditure of one kilowatt for one hour. 1 kWh = 3,413 Btu.

Landfill Gas: Gas that is generated by decomposition of organic material at landfill disposal sites.

Lipid: Any of various substances that are soluble in non-polar organic solvents (as chloroform and ether), that with proteins and carbohydrates constitute the principal structural components of living cells, and that include fats, waxes, phosphatides, cerebrosides, and related and derived compounds.

Lignin: An amorphous polymer related to cellulose that, together with cellulose, forms the cell walls of woody plants and acts as the bonding agent between cells.

Life Cycle Assessment (LCA): LCA is an internationally recognized assessment model of a product's impact on energy, economic, and environmental values. LCA extends from "cradle-to-grave": from material acquisition and production, through manufacturing, product use and maintenance, and finally, through the end of the product's life in disposal or recycling. The LCA is particularly useful in ensuring that benefits derived in one area do not shift the impact burden to other places within a product's life cycle.

Methane: An odorless, colorless, flammable gas with the formula CH_4 that is the primary constituent of natural gas.

Municipal Solid Waste (MSW): Garbage. Refuse includes residential, commercial, and institutional wastes and includes organic matter, metal, glass, plastic, and a variety of inorganic matter.

Organic Compounds: Compounds that contain carbon chemically bound to hydrogen. They often contain other elements (particularly O, N, halogens, or S).

Precommercial Thinning: Thinning for timber stand improvement purposes, generally in young, densely stocked stands.

Pyrolysis: The thermal decomposition of biomass at high temperatures (greater than 400°F, or 200°C) in the absence of air. The end product of pyrolysis is a mixture of solids (char), liquids (oxygenated oils), and gases (methane, carbon monoxide, and carbon dioxide) with proportions determined by operating temperature, pressure, oxygen content, and other conditions.

Quad: One quadrillion Btu (10 15 Btu). An energy equivalent to approximately 172 million barrels of oil.

Residential Sector: An energy-consuming sector that consists of living quarters for private households. The residential sector excludes institutional living quarters.

Residue: Unused solid or liquid by-products of a process.

Rural: Of or relating to the small cities, towns, or remote communities in or near agricultural areas.

Sewage: The wastewater from domestic, commercial, and industrial sources carried by sewers.

Silviculture: A branch of forestry dealing with the development and care of forests.

Syngas: A syntheses gas produced through gasification of biomass. Syngas is similar to natural gas and can be cleaned and conditioned to form a feedstock for production of methanol.

Therm: A unit of energy equal to 100,000 Btus; used primarily for natural gas.

Transportation Sector: An energy-consuming sector that consists of all vehicles whose primary purpose is transporting people and/or goods from one physical location to another. Vehicles whose primary purpose is not transportation (e.g., construction cranes and bulldozers, farming vehicles, and warehouse tractors and forklifts) are classified in the sector of their primary use.

Urban: Of, relating to, characteristic of, or constituting a city, usually of some size.

Chapter 8

Small Wind Power
Energy Roadmap

A Glimpse of the Future

By 2020...
- Small wind turbine sales will contribute 3% or 50,000 MW to America's electric supply.
- The small wind industry will grow to a billion dollar industry, employing over 10,000 people in manufacturing, sales, installation, and support.
- Homes, farms, and communities will install rooftop wind turbines that will help power electric loads and send power back to the utility grid.
- Small wind turbines will be considered a "home appliance," and could be purchased at local home improvement stores nationwide.
- Installed wind turbines will see a 50-year life based on reliability improvements.

INTRODUCTION

The purpose of this chapter is to outline a technology roadmap for the small wind power industry. Small wind turbines are defined as having a generating capacity up to 100 kilowatts (kW), which is about a 60-foot rotor diameter. The roadmap presents a framework that can serve to develop strategic plans for and investment in this technology and business.

Like similar roadmaps in this collection, this roadmap identifies a vision for the small wind industry, barriers to that vision, and appropriate near-term, mid-term, and long-term actions to address these barriers. The roadmap is intended to help guide government and corporate policy towards the overall goal of making small wind a significant contributor to America's domestic energy supply.

THE SMALL WIND INDUSTRY VISION

Vision for Small Wind Power Future

Our vision is to make small wind turbine technology a significant contributor to America's clean energy supply portfolio by providing consumers with an affordable renewable energy option for their homes and businesses and to make wind energy a significant contributor to improving the quality of life and economic opportunities of people in developing nations worldwide through electrification.

Key elements of our vision include:

1. Enhance America's energy diversity and security
2. Increase competition in electric markets by giving consumers the choice of a clean power source.
3. Develop small wind turbines as a household energy appliance and business tool (by lowering competitive energy costs).
4. Build an industry to meet the explosive growth potential.
5. Contribute to rural infrastructure development worldwide.

STATE OF THE SMALL WIND TURBINE *INDUSTRY*

The modern industry for small wind turbines was born in the energy crisis of the 1970s. Responding to the crisis, consumers turned to restored vintage designs from the 1930s, to newly manufactured machines based on the old designs, and to new wind turbine technologies developed to meet modern needs. Most of these turbines were connected to the utility grid. This surge in the U.S. small wind turbine industry, fueled by federal energy tax credits, state incentives, and high electricity prices, peaked in 1983.

Then energy prices fell, federal energy tax credits expired, and state incentives gradually fell by the wayside. By 1986, the people who still wanted small wind turbines were interested in stand-alone or off-grid applications for remote homes. While serving this smaller domestic market, U.S. manufacturers expanded their efforts in markets overseas.

Since 1999, electricity prices have been rising again. People are once again concerned about the security of our energy supplies and the centralized generating facilities that rely on those sources of energy. And some people want independence from electric utilities. There is also a steadily increasing concern about global warming.

State governments, under utility restructuring, have enacted significant incentive programs that buy down the initial cost of small wind turbine systems, thereby tunneling through the cost barrier. These incentives are funded through system benefit charge programs that are significant—totaling $3.5 billion in 2001 for programs that include incentives for small wind turbines. All these factors have increased interest in small wind turbines connected to the utility grid. Meanwhile, the U.S. industry continues to dominate the overseas market for small wind turbines.

In 2001, annual sales of the U.S. small wind turbine industry were estimated to be 13,400 turbines valued at about $20 million.[1] While this is about the same level as sales in the early 1980s, it is only about 2% of the value of sales of large wind turbines in the United States.[2] The success of the large wind turbine industry shows the impact of sustained, substantial support from government programs and policies (both at home and abroad). Support such as federal and state tax credits was discontinued in the mid-1980's for small wind systems. This led to a significant shrinking of the industry and a loss of momentum in technology and market development.

There are several good reasons why it is time for a combined effort from government and industry to increase the contribution of small wind turbines to our generation mix. First, there is the potential for real contribution to our energy supply. We project that small wind turbines could contribute 3% of U.S. electrical consumption by 2020.

Second, small-wind technology is a home-grown industry. While foreign companies dominate the market for other renewable energy technologies, the U.S. small wind turbine industry is the leader in markets at home and abroad.

Third, the market for small wind turbines also fuels companion industries, including those that market composite products, steel, towers, power electronic equipment, and construction projects.

Fourth, while producing energy, small wind turbines produce no environmental emissions.

Fifth, small wind turbines help meet the national need for energy diversification and national security.

And finally, the American public overwhelmingly supports the expansion of renewable energy, and they stand to benefit from more choices about where their energy comes from.

Recently, the market for small wind turbines has been growing 40% per year. As we discuss later, the potential market for residential and business applications of small wind turbines is tremendous because it is clear that the turbines work and that people want them. However to realize our vision, significant challenges lie ahead in the market, policy, and technology areas. We must identify these challenges and barriers; we must devise actions to overcome these barriers; and finally, we must devise a strategy of public/private cooperation to complete these actions and reach our goals. In the end, we expect to deliver small wind turbine products that people desire and that they can afford, allowing individuals to contribute to our energy security.

STATE OF SMALL WIND TURBINE *TECHNOLOGY*

The U.S. small wind turbine industry offers a wide assortment of products for various applications and environments. Machines range in size from those that generate 400 watts (W) of electricity for specific small loads such as battery charging for sailboats and small cabins, to 3-15 kilowatt (kW) systems for a home, to those that generate up to 100 kW of electricity for large loads such as a small commercial operation.

Small wind turbines can operate effectively in most of the rural areas of the United States. In fact, about 60% of the United States has enough wind for small turbines to generate electricity. Today's small turbines have been designed for high reliability with only two or three moving parts and therefore have relatively low maintenance requirements.

Thanks to continuous development within the industry and in collaboration with the U.S. Department of Energy (DOE) small wind turbine projects, several new features are in development for incorporation into commercial turbines. Advanced airfoils, super-magnet generators, smart power electronics, very tall towers, and low-noise features will help reduce the cost of electricity and increase the acceptability of this technology.

Small wind technology has been improving since the 1970s. However, it is still generally acknowledged that more work is needed to improve operating reliability, eliminate noise concerns, and lower manufacturing and installation costs. There is much to be done both to incorporate the technologies currently under development and to enhance manufacturing. As an example of the cost reductions that are possible, the industry estimates that high-volume manufacturing alone could reduce costs 15-30%.

Figure 8-1. Small 10 kW Wind Turbine for Typical Home Use

TECHNOLOGY OPPORTUNITIES

Modern small wind turbines are not like our grandparents' wind generators from the 1920s and 1930s. Today's small turbines borrow from aerospace technologies with sophisticated, yet simple, designs that allow them to operate reliably for up to a decade or longer without maintenance. Current products are designed for operational lives of 20 to 30 years, and they have withstood everything, short of a direct hit from a tornado, that Mother Nature can throw at them. As small wind turbine technology has matured, the products have become mechanically simpler and more robust.

Responding to more active markets in the last few years, the small wind turbine industry has increasingly adopted advanced component technologies and state-of-the-art design tools such as three-dimensional solid modeling and computational fluid dynamics. Technologies such as unique high-efficiency airfoils, neodymium-iron-boron "super-magnet" generators, pultruded FRP blades, graphite-filled injection molded plastic blades, special purpose power electronics, and tilt-up tower designs have both lowered costs and increased efficiency. The long-term vision of the industry is to produce small wind turbines that are accepted as common household appliances in the same way that heating and air-conditioning systems are today. By virtue of their compelling economics, these new turbines will achieve high market penetration in areas with lower

People, however, do not tend to live where the wind howls, so achieving high market penetration rates will require small wind turbines that are specially designed to work effectively in low wind resource areas. These turbines of the future will need to have relatively larger rotors to capture more energy. But they cannot sacrifice robustness because even areas with low average wind speeds experience severe weather. The new turbines must be extremely quiet, so that they are seldom heard above the local background noise. They must be able to operate for 10 to 15 years between inspections and/or preventive maintenance, and they must offer a reasonable expectation of a 30- to 60-year operating life. Most important of all, the small wind turbines of the future must be affordable without significant subsidies.

Achieving these goals will require further advances in small wind turbine technology, major improvements in small turbine manufacturing, and more efficient installation techniques. The U.S. Department of Energy (DOE) and the National Renewable Energy Laboratory (NREL)

have critical roles to play in accelerating the development and adoption of new small wind turbine technology and manufacturing techniques. A close working relationship between DOE/NREL and the small turbine industry is important today, and it will become increasingly important as international competition heats up over the next five to ten years. All parties need to realize that large wind turbines are now in their seventh or eighth generation of technology development, while small wind turbines are only in their second or third.

For its part, the industry is striving to reduce the cost of electricity generated by small wind turbines. In 2002, typical 5- to 15-kW residential wind turbines cost about $3,500 per installed kilowatt. These turbines produce about 1,200 kWh per year of electricity per kilowatt of capacity in an area with a DOE class "2" wind resource. By 2020, the industry hopes to have lowered the installed cost to between $1,200 and $1,800 per kilowatt (smaller systems being relatively more expensive) and to have raised the productivity level to 1,800 kWh per installed kilowatt. If these goals are met, the 30-year life cycle cost of energy will be in the range of $0.04 to $0.05/kWh, lower than virtually all residential electric rates in the country today.

To further enhance the attractiveness of small wind turbines to consumers, there is also a need for meaningful, appropriate, and cost-effective standards and a certification program for them. Some new entrants to the industry have significantly underestimated the engineering rigor and expense required to deliver a reliable small wind turbine product. And, in light of recurring instances of exaggerated claims, consumers have had trouble sorting out reasonable from unreasonable claims of performance. The standards and certification programs that exist for large wind turbines are not appropriate for small wind turbines. Appropriate standards for small wind turbines are under development by the international industry and by research institutions. However, the U.S. industry and DOE must also work to ensure that related standards, such as electrical grid interconnection standards, are justified and do not unduly raise the costs of owning a small wind turbine.

The industry believes that research cooperation between the private and public sectors is strategically important if the U.S. industry is to maintain its leadership position. The engineering challenges presented by the interlocking disciplines of aerodynamics, structures, controls, electrical conversion, electronics, and corrosion prevention are formidable. There are also a number of generic technology opportunities that are not likely to be fully explored by the private sector alone. Government and industry must work together to build a better small wind turbine.

To assist industry in addressing technology barriers, four models of government/industry collaboration are employed.

1. Research conducted at national laboratories and universities with input from members of the industry.
2. Applied research projects conducted at the facilities of small wind turbine companies with support from the government through competitive procurement.
3. Applied research projects involving companies, universities, and national laboratories.
4. Privately funded research and development.

The opportunities offered by improved technology can be achieved through the cooperative activities discussed in this roadmap for the small wind turbine industry.

Work by industry members, research institutes, state and local governments, and DOE can help increase the contribution of small wind turbines to the electricity generation mix.

MARKET POTENTIAL

In 2001, we estimate that 13,400 small wind turbines were manufactured in the United States. More than 50% of these were exported. We believe that both the domestic and foreign market for small wind turbines will continue to grow. This roadmap is designed to accelerate this growth to its maximum potential.

U.S. Markets

We estimate that small wind turbines have the potential to contribute up to 8% of U.S. electrical demand in 2020. Our industry goal is to install turbines that will generate at least 3% of U.S. electrical demand in 2020 or 6-8% of residential electricity demand. This will require small wind turbines installed with a total generating capacity of 50,000 MW.

The most recent public market study for small wind generators was the A. D. Little study sponsored by DOE in 1981.[1] That report (the ADL study) projected a market potential of 3.8 million small wind systems installed in grid-connected applications. If the average generating capacity of these systems were 10 kW, then the potential contribution to the nation's generation mix would be 38,000 MW.

The largest potential market for small wind turbines is for homeowners in rural areas where wind-generated electricity can reduce utility bills. In 1998, American homes used 1.1 trillion kWh or 35% of total electricity sales. While some wind turbines may be installed when a new home is built, most market opportunities will be for installations at existing homes. In 2020, there will be approximately 43 million homes with 1/2 acre or more of land. Of these homes, we estimate 65% will be prevented from using small wind technology because the wind resource is not sufficient, because of restrictive zoning and covenants, or because of proximity to airports or other sensitive areas. This will leave 15.1 million homes with the potential to install a small wind turbine. If each of these homes installed a 7.5-kW machine, the total contribution to generating capacity would be 113,000 MW.

When combined, other markets for small wind turbines in the United States offer significant opportunities to expand electric generation capacity. For example, about two million medium-sized commercial buildings are candidates for small wind turbines of 10 to 100 kW. In addition, public facilities such as schools and government buildings could also use small wind turbines at suitable sites.

Another distributed generation market sector includes industrial and commercial customers who are connected to utility grid and may have back-up generation requirements, which could easily be integrated with a larger small turbine. Since the utility rate structure is typically different from the residential market (e.g. demand charges), further study is needed to specifically define this market.

Where the utility grid is not available, stand-alone or hybrid systems could provide electricity for homes, communities, water pumping, and telecommunications services. The Energy Information Administration (EIA) estimates that there are 200,000 off-grid homes in the U.S. This is already a very active market for small wind systems.

There are also a number of off-grid communities that are remote, isolated, and produce their electricity with diesel or gasoline generators. Alaska, for example, has 91 villages powered by diesel generators, serving a population of about 42,000 people. In addition, several hundred miscellaneous remote facilities are powered by diesel generators ranging in size from 2 to 250 kW.

In addition, water pumping for livestock and off-grid facilities is still a sizable market. In the early part of the 20th century, the United States had about three million mechanical windmills in operation, supplying water for homes, farms, and livestock. Today, there are also new wind-electric water pumping systems for which the turbine can be located where there is good exposure to the wind, and it does not have to be located near the well and pump. However, for low wind

Figure 8-4. Wind Powered Water Pump in Rural Area

speed sites, the mechanical water pumper still offers more economic water pumping.

Deregulation of the telecommunications industry and the rapid growth in wireless systems has spawned growth in the development of remote broadcast facilities. The preferential method of powering these facilities is hybrid systems that combine generation from solar, wind, and diesel systems.

These other markets could contribute up to 25,000 MW of generating capacity by 2020. From this discussion, we conclude that the total installed capacity for small wind turbines in 2020 could be 140,000 MW across all markets. However, the goal of the AWEA Small Wind Turbine Committee is to install 50,000 MW of small wind turbines by 2020.

According to figures taken from the most recent EIA documents, the total generating capacity in the U.S. in 1999 was approximately 745,000 MW. According to the *EIA Annual Energy Outlook 2001*, the projection for 2020 is 1,060,000 MW of generating capacity and 4,804 billion kWh in demand.

Although the domestic potential for small wind generating capacity is estimated at 140,000 MW in 2020, we do not believe that this is a realistic goal. The limitation we see is market growth, not manufacturing capacity or sales and support infrastructure. The growth of small wind turbine markets, even with attractive incentives and favorable policies, will not match the pace of market penetration of other common household electrical devices with lower price tags and easier implementations.

Our goal of 50,000 MW of small wind capacity by 2020 is aggressive but achievable given the right public policy environment, particularly over the next ten years. Fifty gigawatts (50,000 MW) of small wind turbines in 2020 would produce an estimated 132 billion kWh of clean electricity per year, or approximately 3% of projected total U.S. demand. At this level of capacity, small wind systems would be providing 6-8% of residential sector electrical demand. The EIA *Annual Energy Outlook* document forecasts that the residential electric sector demand will be 1,701 billion kWh in 2020.

Growing the domestic market from its current installed capacity of 15-18 MW to 50,000 MW in 2020 would require a doubling of the market each year for several years and then require sustained sales growth in the range of 50-55% per year. In this scenario, the domestic small wind turbine industry would reach annual sales of $1 billion and employ approximately 10,000 people in 2020. And other off-grid sites present a sizable market that could contribute up to 25,000 MW of generating capacity by 2020.

Export Market

U.S. manufacturers of small wind turbines currently export more than 50% of their production and have a leading share of the world market for this technology. The foreign market for grid-connected wind turbines is fueled by electricity prices more than double those faced by U.S. consumers. In addition, it has been estimated that about 2 billion people in the world do not have access to electricity for domestic, agricultural, or commercial uses. The traditional method of providing electricity by extending the distribution grid has proved to be expensive and poorly suited to the low consumption levels of communities in developing nations. And the number of homes without electricity is increasing because the birthrate is outpacing the electrification rate.

Small-scale renewable energy systems (wind, micro-hydro, and solar) are often less expensive to install than line extensions. Small turbines are less expensive to operate and produce much less carbon dioxide per kilowatt hour than diesel generators do. Small wind systems can be used to electrify single homes (<500 W) or villages (<50 kW). There are also a myriad of special uses of wind electricity, such as making ice for coastal fishing villages, charging batteries for distribution to single homes, and purifying water for drinking.

Developing countries have a high potential demand for small wind systems because they normally do not have major electrical power plants serving rural areas. However, the people are usually too poor to buy small wind systems and need financial assistance from their government in order to afford them. This assistance is, today, almost exclusively directed to subsidizing extension of the grid and installing diesel generators.

BARRIERS TO MARKET DEVELOPMENT

Identifying and prioritizing barriers is an important prerequisite to industry planning. Different companies and other stakeholders often have differing views on the most important barriers, and there is a natural tendency to focus on short-term challenges. Reaching consensus on the barriers required a number of meetings and sizeable investment in time, but this consensus provides the foundation for plotting the path to a billion-dollar industry.

Outlined below are technology, market, and policy barriers for the near-term (0-3 years), mid-term (3-10 years), and long-term (10+ years) time frames.

Near-Term (0-3 Years)
Near-Term Technology Barriers

- High costs of wind turbines. Although people want small wind turbines, most find the price is too high. Increasing production rates and advancing the technology can reduce system costs. Advances in technology, such as new airfoils for blades, super-magnet generators, and power electronics can make small wind turbines cheaper to build, more productive, and more reliable. The cost to consumers can be reduced with incentives.

- Insufficient product reliability. It is a substantial challenge to design, manufacture, and install small wind turbines that are low in cost and yet rugged enough to withstand 20 to 30 years of operation in weather that is often severe. Small wind turbine technology development is both art and science. The true measure of a new design is often not known until several years of operation at dozens of sites. At present, there is no way to effectively duplicate the wear and tear of the real world during the product development stage. As a result, reliability has historically been the Achilles heel for small wind turbine technology.

Near-Term Market Barriers

- Lack of effective standards. As the domestic market continues to expand, consumers need protection from unscrupulous suppliers chasing the latest trend in search of wealth. Further, responsible small turbine manufacturers need a baseline for establishing turbine performance and credibility. There are existing international safety and draft national performance standards for small turbines that could be used by state or national incentive programs. The issue is the cost to meet the testing and documentation requirements of these standards. The existing design standards are primarily related to structural safety and do not directly address issues of reliability, durability, and longevity. There is no equivalent to Consumer Reports in the small wind industry.

- <u>Low visibility of the industry and technology</u>. There are relatively few small wind turbine installations, so people just do not see small wind turbines very often. In addition, the companies of the small wind turbine industry have limited resources and capabilities to promote the technology. Unlike the solar industry, there are no Fortune 500 companies involved at this time in the small wind industry. Adding to the issue, DOE has focused the majority of its resources and publicity on solar technologies.

- <u>Misconceptions about the wind resource</u>. The attention given to wind farm developments in high-wind areas has convinced some people that they must have an exceptional wind resource in order to benefit from wind technology. However, small wind turbines are designed to operate effectively in the lower wind speed areas where most people live and work. DOE wind maps have inadvertently exacerbated the problem by classifying wind regimes according to their potential for wind farm development.

Near-Term Policy Barriers
- <u>Lack of federal incentives</u>. For small wind turbines, there has been no federal tax incentive or deployment program since 1985. The resulting lower level of business activity has affected industry expenditures on research and development and has slowed the pace of innovation. DOE has supported sporadic initiatives for research and development on small wind turbines since 1985. However, this level of investment by the federal government has not been sufficient to realize the potential for cost reductions or leveraged deployment.

- <u>Restrictive zoning</u>. Most local jurisdictions limit the height of structures in residential and sometimes other zones to 35 feet. This restriction was developed nearly 100 years ago to ensure that the height of structures would not exceed the capability of fire fighting equipment to pump water. Today, this height limit is a significant obstacle to siting small wind turbines. To make effective use of the wind, small turbine towers must be at least 60 feet high and well above obstacles (such as trees) in their vicinity. Wind speed increases with height above the ground. Turbulence, the disruption of the wind flow around obstacles, increases near the ground and reduces energy output of small wind turbines. Putting a small wind turbine on a short tower is like putting a solar panel in the shade. For many residential applications, systems of 5 to 15 kW, turbines need to be on towers from 80 to 120 feet tall. The 35-foot height restriction causes unnecessary expense and delay when getting a building permit and opens up opportunities for neighbors to oppose the permit because of either legitimate concerns or underlying historical animosities.

- NIMBY and environmental concerns. Because there are few installations of small wind turbines, the neighbors of people planning to install a small wind turbine system and the local zoning boards that must approve permits typically do not have a fair basis for understanding the noise, visual, and other impact of the turbines. They fear the worst and act accordingly. Consumers who need approval from zoning boards often run into objections presented by concerned neighbors. Dealing with these objections can demand considerable time and expense.

- Excessive interconnection requirements and unequal billing policies. Even though the federal Public Utility Regulatory Policies Act (PURPA section 210), gives all Americans the right to inter-connect small wind turbines and to receive payment for excess electricity production, the policies of many utilities discourage the use of these systems. Many utilities have limited experience with customer-owned generation. They may use the same process for approving 500-MW gas turbine cogeneration facilities and 10-kW residential wind installations.

It sometimes takes more hours of labor by the customer and the wind turbine vendor to gain the approval for interconnection than it took to build and install the wind turbine itself. Weak or uninterested public utility commissions can allow utilities to effectively thwart the federal rights provided under PURPA. Interconnection standards that have emerged in the last five years have required small wind turbines to deliver to the utility grid power that is of higher quality than the power delivered by the utility to its customers.

Meeting these excessive standards increases the cost of the wind generating systems. In more than 35 states, there is a policy called net metering, in which consumers receive the retail rate for electricity they generate in excess of their consumption. But even under net metering, there is a great variation from state to state and utility to utility concerning the accounting periods, capacity limits, limits to participation, and cost/benefit of net excess generation. This creates uncertainty in the marketplace.

- <u>Underevaluation of green power</u>. No economic credit is given for having a nonpolluting energy source.

- <u>Disincentive in tax code</u>. Companies that buy fuel or electricity can deduct these costs as business expenses, reducing their reported profits and their tax liability. Companies investing in energy-producing equipment, on the other hand, must capitalize the investment, increasing reported profits and tax liabilities. The accelerated depreciation schedule for wind energy investments helps, but does not level the playing field.

Mid-Term Barriers
Mid-Term Technology Barriers
- <u>Turbine productivity hampered by power electronics issues</u>. Because small turbines operating at variable speed produce variable frequency and variable voltage output, power electronic converters are used to modify the wild AC into standard 60-cycle AC. The old inverter technology was too unreliable. The new inverter technology is too expensive. The industry needs lower cost, reliable power electronics.

- <u>Domestic market requirement—quiet operation</u>. High-growth domestic markets demand quieter wind turbines, especially when turbines are sited in residential neighborhoods. Turbine noise can be caused by gearboxes, blade shape, tower shadow, etc. Small turbines operate at high RPM and tend to spin even if they are furled (pointed out of the wind); there needs to be a way to make them extremely quiet under all conditions.

- <u>Long-term reliability</u>. As mentioned above, reliability is a long-term issue that is difficult to predict because the wind is such a variable environment for a piece of machinery. How a turbine withstands the long-term effects of the wind is often known only after several years of operation at dozens of sites. There needs to be a way to effectively duplicate the wear and tear of the real world during the product development stage.

- <u>Need for better technology tools</u>. Many small turbines use a passive over-speed control such as furling. In furling, the force of the wind turns the rotor sideways, just as farm water-pumping windmills have done for 100 years. So far, no computer codes have been able to predict the performance or assist in the design of furling mechanisms. This means such designs need to be performed empirically, raising development costs. Better computer codes are needed to help reduce design costs.

Mid-Term Market Barriers
- <u>Insufficient capitalization</u>. U.S. small turbine manufacturers are entrepreneurs who have a deep dedication to renewables, but who also have limited resources. Their businesses lack the capitalization to effectively promote mass markets, exploit design-to-cost technology options, or provide forward pricing to accelerate market adoption. While the solar industry has consumed billions in investment over the last two decades, the small wind turbine industry has attracted far less capital.

- <u>Complicated financial impact</u>. For consumers, purchases like small wind turbines that have longer lifetimes are more difficult to understand financially. Consumers generally look at monthly cash flow or direct out-of-pocket expenses and rarely consider life-cycle costs. They also do not tend to consider tax consequences fully. This makes it difficult for them to compare small wind turbines with other potential investments.

Mid-Term Policy Barriers
- <u>Need for more state-based incentives</u>. Currently, four states (California, Illinois, New Jersey, and Rhode Island) offer substantial rebate or buy-down programs to promote the installation of renewable energy equipment. Rebates typically range from 50 to 60% of the installed small wind system cost, resulting in significant savings. Other states, such as Wisconsin, offer production-based incentives. If more states offered incentives, the sales and production volume for small wind turbines would increase.

- Need for sustained national incentives. The lack of federal incentives slows the pace of industry growth to meet the market demand. Other traditional energy source technologies are being subsidized, and as noted above, federal tax policies actually encourage the use of fossil fuel and utility power.

- Need for national models for net metering and zoning rules. There are too many state, county, and city jurisdictions for the wind industry to address the policy needs of each. For example, in California there are 538 counties and incorporated cities. A 2001 industry study estimated that addressing the zoning barriers with new ordinances in each jurisdiction would cost more than $20 million and would require more than 200 person-years of effort.

- Lack of interconnection standards. Few states have standardized interconnection requirements, and many public utility commissions give utilities broad discretion on policies towards customer-owned generation. Some utilities have used this freedom to discourage competition through excessive requirements for equipment, special tests, and additional insurance. There is a need for a national interconnection standard developed by a technically recognized body.

GOALS AND ACTIONS TO REMOVE BARRIERS

This roadmap of the U.S. small wind turbine industry identifies the background, status, and potential of the market for small wind turbines. The roadmap points to technology, market, and policy goals and the specific actions necessary to meet these goals. Many groups will need to participate in the activities described here in order to realize our vision of U.S. small wind turbines as a significant contributor to America's energy supply portfolio.

These high-priority goals have been identified by the American Wind Energy Association (AWEA) Small Wind Turbine Committee to overcome the barriers identified in the previous sections. Goals and actions are listed that can be taken by the industry working in concert with federal, state, and local governments to meet these goals. Progress on these items in the near-term, mid-term, and long-term time frames will help make the vision of this roadmap a reality.

Near-Term (0-3 Years)

Near Term Technology Goals and Actions

- Reduce cost of energy resulting from turbines that operate in low-wind regimes by developing technologies for low-cost towers and low-wind rotors.

- Reduce turbine cost through improvement of the performance and efficiency of small wind turbines through cost-shared RD&D and advanced technology development.

- Reduce tower and installation costs by developing advanced, lower-cost foundation or anchoring systems for towers and automated processes for tower fabrication.

- Improved turbine reliability by developing test methods for reliability issues like "extreme events," and gathering multi-year data on turbine performance, reliability, operation, and maintenance.

- Increase participation of small wind turbines as a technology option in domestic government programs by working with the Federal Energy Management Program to develop small wind projects at federal facilities and promote small wind turbines for homeland security and other military operations.

- Reduced manufacturing costs by increasing the volume of production and improved manufacturing techniques.

- Develop equipment and processes for mass production of small wind turbine systems.

Near-Term Market Goals and Actions

- Develop of additional efficient and effective standards to address reliability, durability, longevity, noise, and power performance.

- Develop stronger, certified distribution channels and support, particularly through generic installation and maintenance training programs for small wind turbines that include certification standards.

- Develop of a consumer-friendly performance rating system by updating and reconciling the AWEA performance standard with the IEC 61400-12 for small wind turbines.

Figure 8-3. Example 50 kW Wind Turbine with 15m Rotor Diameter

- Increase visibility and credibility of small wind turbines by encouraging DOE to give small wind greater visibility through policy
  incentives, studies, and speeches and by publishing in cornerstone
  magazines such as Scientific American and Popular Science to highlight small wind as an important technology. Also, create a significant federal deployment initiative for small wind turbines to prime
  important markets.

- Complete high-definition wind mapping for all states and for international markets.

Near-Term Policy Goals and Actions
- Support national and state policies to promote market development, such as tax credits, regulations, incentive programs, rebates, etc.

- Remove systemic height restrictions normally found in residential locations

- Provide more information to address aesthetic, noise, and environmental concerns

- Remove interconnection barriers by providing information to utilities and public utility commissions that will help remove unnecessary and expensive requirements in the areas of power quality, safety, and performance standards; also participate in the development of a national interconnection standard.

- Equitable utility billing and interconnection cost policies by promoting net metering and simplified interconnection contracts.

- Provide credit for green attributes of wind power.

- Reduce or eliminate disincentives to investment in small wind turbines (such as disincentives in the tax code, sales taxes, property taxes, etc.).

Mid-Term (4-10 Years)
Mid-Term Technology Goals and Actions
- Develop improved power electronics through cost-shared company research on power electronics equipment and applied research on generic power electronics issues.

- Reduce noise produced by small wind turbines through a noise measurement and reporting standard for small wind turbines and noise reduction research and development.

- Create higher-definition performance predictions to help customers in site selection; in particular, develop a web-based performance prediction capability based on high-definition wind maps.

- Improve the reliability and durability of small wind turbines through improved life-cycle testing protocols and analytical methods for small wind turbines and durability and reliability testing for environmental extremes.

- Reduce maintenance requirements of small wind turbines through RD&D projects.

- Develop enhanced analytical tools for small wind turbine design, particularly for rotor aerodynamics and dynamics that are unique to small wind turbines.

Mid-Term Market Goals and Actions
- Better determine the small wind market (to be used for business planning) and characterize the export potential of U.S. technology.

- Increase customer options for purchase and financing small wind turbines by expanding the availability of plug-and-play systems suitable for mass marketing.

- Increase the number of products available (models and size range) for different market segments.

- Incorporate environmental benefits into the value of wind turbines through green tags markets for distributed generation.

- Increase outreach and education on small wind turbines through state-specific consumer guides for small wind turbines and wind power for school programs.

- Increase participation of small wind as a technology option in international development projects.

Mid-Term Policy Goals and Actions
- Support state policies for small wind incentives

- Disseminate information on market barriers, for example streamline information on zoning regulations, interconnection agreements, and net metering to guide development of state incentives.

Long-Term Action (11+ Years)
Long-Term Technology Goals and Actions
- Develop hydrogen-based systems by establishing a link with other hybrid power technologies such as micro gas turbines, PV panels, diesel and other fuel generators, and any new power generating technologies that may develop.

- Establish links with storage and other power technologies such as hydrogen generation and storage, batteries, natural gas sequestration, and any new storage technologies that might develop.

Long-Term Market Goals and Actions
- Stimulate the micro-power revolution by developing a strategy with outreach materials to address the micro-power market.

Long-Term Policy Goals and Actions
- Develop strategies to deliver high-level electric service to rural customers by working with rural electric cooperatives to devise incentives for rural customers and to streamline the interconnection process.

STRATEGY

> "Many more people would buy small wind systems if they were cheaper. But, we can't make them cheaper unless many more people buy them."
>
> *David Blittersdorf, President, AWEA.*

All emerging industries and products have faced this basic "chicken and egg" dilemma. Some never made it over the hump and eventually faded from the marketplace. Some got over the hump by virtue of massive corporate investment (in R&D and forward pricing), and others made it into the mainstream by steadily improving cost and functionality. The small wind industry is composed of innovative, small, struggling firms that lack the resources to develop mass-production designs and build large factories in anticipation of quantum increases in sales. Steady improvement is the most likely route for firms offering small wind turbines. In Washington, this is sometimes referred to as sustained orderly development.

Sustained orderly development is the recognition that there are not likely to be "silver bullets" that will radically transform the glide slope of market penetration for small wind turbines. Steady improvements in the products and sustained public sector support offer the best hope of delivering the industry's vision of a "new home appliance" and making small wind turbines a notable contributor to our national energy supply. Although economics are a barrier, the exciting thing about residential and small business markets is that once the numbers work for one home, they work for 10,000 homes. This gives the small wind turbine industry explosive growth potential.

Public Policy

A primary element of sustained orderly development for small wind turbine technology is related to public policy. Smart subsidies, which enable customers to tunnel through the cost barriers, are critically important in aggregating demand. Federal and state subsidies and incentives need to dovetail for a total package that provides enough stimulus to move the market. Removing institutional market barriers, such as tower height restrictions and undue inter-connection costs, is also an important part of the recipe.

In addition to efforts by the U.S. government, some states have good policy environments for small wind turbines. Many of these states offer rebate or buy-down programs that are typically funded with system benefit charges assessed to retail electricity sales. Funds generated by these system benefit charges can be designated by legislatures to subsidize renewable energy projects and promote industry development.

In 2002, four states offered these rebate incentives for small wind turbines: California, Illinois, Rhode Island, and New Jersey. The requirements for specific turbines are determined on a state-by-state basis, and typically these incentives are only for small turbines that are connected to the grid. Other states offer tax credits, sales or property tax incentives, and net metering for small turbine owners. As the number of states offering stimulus packages for small turbines increases, there will be further sustained orderly development of the market.

In order to meet the market goals of the AWEA Small Wind Turbine Committee, more incentives are needed. An additional two or three states per year need to implement stimulus programs for small wind turbines over the next six to ten years. It is also critically important that the federal government does its share with a significant (25 to 40%) tax credit or rebate program.

Research, Development, and Demonstration

Federal, and to a lesser extent, state R&D programs need to provide greater resources for small wind turbine technology development, and the industry must continue to devote significant resources to product development. The DOE research program on wind energy cannot be effective if it concentrates solely on long-range, high-risk R&D. Instead, the federal R&D program must support development of advanced products and components and attack other cost drivers such as technology for manufacturing and installation, outreach for consumer education, support for policy and market transformation, and work on generic basic technology.

DOE now has a wind energy R&D budget of about $40 million, of which about $3 million, or 8%, is spent on programs to develop small wind turbines. As technology for large wind turbines "graduates" and requires less public sector support for the next few years, spending on small wind technology should significantly increase.

Small Wind Turbine Industry

The small turbine industry must be open to new entrants and should not foreclose any investment options. Private capital will remain the cornerstone of the substantial investments that will be necessary if this is to become a billion-dollar industry. Likewise, the industry should pursue all viable market diffusion models, from full-service dealerships to "big box" chain stores to "Dell-like" direct sales programs.

Cooperative Strategy

The members of the industry will continue to develop products and processes to meet market demand. DOE, national laboratories, test centers, and universities will work with industry partners to conduct basic and applied research, development, and testing to improve small wind technology. Organizations representing utilities, standards-making bodies, regulatory agencies, and every level of government will participate in market and policy actions to remove barriers. Consumers will express their will by seeking out this environmentally friendly technology in spite of the barriers that still exist.

To address policy barriers, industry and government entities work together to propose, review, advocate, and help implement policies that support development of small wind technology while safeguarding the public interest. The industry also lobbies at the national, state, and local level for policies that remove barriers and compensate for subsidies to other energy technologies.

Table 8-1 summarizes the wide-ranging discussions of the AWEA Small Wind Turbine Committee about the timing of actions to overcome barriers and reach the goals described in this roadmap.

CONCLUSION

There is great potential for small-scale wind power in the U.S. and abroad. This roadmap identifies the barriers, challenges, and opportunities for this growing industry. With appropriate incentives and technology advancement, the small wind turbine industry will be prepared to advance into markets that are now unavailable.

This roadmap represents an activity that was mostly driven by industry. The chapter was derived from the report *The U.S. Small Wind Turbine Industry Roadmap: A 20 Year Industry Plan for Small wind Turbine Technology* produced by American Wind Energy Association (*www.awea.org*). The roadmap was developed by a dedicated group of representatives of the wind power industry, in conjunction with personnel from the public sector. More details about the content in this roadmap can be found on the AWEA website, including a list of state incentives for wind power development. This roadmap is reprinted here in its abridged form with permission from AWEA.

[**Editor's Note**: This chapter represents an abridged and edited version of the report entitled *The U.S. Small Wind Turbine Industry Roadmap: A 20 Year Industry Plan for Small Wind Turbine Technology*, published by AWEA, 2002. The editor is grateful to AWEA for permission to publish an abridged version of this report.]

References
[1]Bergey, Mike, "The Sleeping Giant: The Role of Small Wind in America's Energy Future," in Proceedings of Windpower 2001. American Wind Energy Association: Washington, DC, June 7, 2001.
[2]Arthur D. Little, Inc., Near-Term High-Potential Counties for SWECS. Solar Energy Research Institute: Golden, CO, BE-9-8282, 1981.
[3]Arthur D. Little, IBID.

Table 8-1. List of Actions by Timeframe

	Technology Actions	Market Actions	Policy Actions
Near-term	• Reduce costs by new turbine development activity for low wind speed sites and new component development for SWT • Research reliability concerns such as lightning, corrosion, bearing lubrication, alternator winding insulation, electronics • Continue focused long-term research unique to SWT - furling, durability, blade aerodynamics, noise, and power electronics • Develop packages with other distributed generation and storage technologies	• Develop nationally recognized standards for participation in stimulus programs • Publish SWT articles in cornerstone magazines such as *Scientific American*, to create more "SWT buzz" • Revise new U.S. wind maps for SWT, 3 m hub height and .25 shear, new legends • Explain turbine micrositing • Provide information to remove misconceptions about the wind resource • Incorporate the value of environmental attributes of small wind into electricity prices	• Develop a strategy to work with state policies for inclusion of small wind • Develop a national policy for an SWT tax credit • Work to eliminate zoning restrictions • Develop model zoning ordinances and blueprint templates of zoning regulations, interconnection agreements, and other policies • Work to reduce excessive interconnection requirements

(Continued)

Table 8-1. (*Continued*)

	Technology Actions	Market Actions	Policy Actions
Mid-term	• Work to improve the reliability and reduce the cost of power electronics • Work to eliminate noise from small turbine designs • Develop consumer-friendly performance predictions • Improve analytical design tools • Continue the development of packages with other distributed generation and storage technologies	• Update national market study • Characterize the export potential for U.S. manufacturers and work with multilateral development programs • Establish consumer-friendly customer financing programs, including lease options • Increase the number of products available (models and size range) for different market segments • Increase outreach and education	• Influence/develop new state and national incentives • Disseminate and expand information on zoning regulations, interconnection agreements, and net metering rules • Develop a more consumer-friendly national interconnection standard
Long-term	• Develop hydrogen-based systems • Develop blackout protection strategies • Establish links with storage and other power technologies	• Stimulate the emerging micro-power revolution, of which SWTs are part	• Develop policies to help deliver higher service levels to rural customers
Crosscutting	• Continually work to reduce cost and improve reliability • Continue to develop standards for reliability, durability, and longevity	• Continue to develop standards for reliability, durability, and longevity	• Continue to develop standards for reliability, durability, and longevity

Chapter 9

Solar Electric Power Technology Roadmap

A Glimpse of the Future

By 2020...

- The domestic photovoltaic industry will provide up to 15% (about 3,200 MW) of U.S. peak electricity generating capacity.
- Cumulative installed capacity in the United States for PV will be about 15 GW.
- Conversion efficiencies will be on the order of 18% to 20% at a cost of less than 50 cents per watt for each module technology.
- The industry as a whole will have annual revenue of $10-$15 billion.
- Production facilities will be largely automated, thus reducing the costs of production by a hundredfold from 2000 values.
- PV systems will have "plug-and-play" capabilities that will tie into the existing grid or homes for electricity production.

INTRODUCTION

The 21st century brings numerous challenges and opportunities that will affect our nation's energy, economic, and environmental security: Global economic and population growth. Technological advances. Utility restructuring. Greater demand for power quality and reliability. Environmental sensitivities. Global warming. These major driving forces underlie the need—and the opportunity—for improving our nation's systems for generating and delivering energy.

Some of these forces have already affected our nation's energy supply and security. For example, the prices of natural gas and oil have increased dramatically and have exhibited considerable volatility. Electricity restructuring in California has caused enormous increases in electricity prices recently, while during the summer of 1999, stress on the transmission and distribution system caused widespread power outages that affected millions of people and thousands of businesses. Meanwhile, our nation is increasingly reliant on imported fossil fuels.

A comprehensive national energy strategy is required to meet these challenges. Renewable energy will be an important element of an energy portfolio that improves our energy security, preserves our environment, and supports economic prosperity. A portfolio of renewable energy technologies could provide a significant fraction of the nation's electricity generation requirements and, in concert with other generation sources, provide more reliable power.

Significant among the renewable technology options is solar-electric power—specifically photovoltaics (PV), which is a semiconductor technology that converts sunlight into electricity. The "photovoltaic effect" produces direct-current (DC) electricity, while using no moving parts, consuming no fuel, and creating no pollution.

Solar-electric power is ideally suited to be a major contributor to an emerging national energy portfolio. The U.S. electrical grid will increasingly rely on distributed energy resources in a competitive market to improve reliability and moderate distribution and transmission costs and on-peak price levels. Distributed power also allows greater customer choice—for example, some consumers place great value on power reliability or clean power, as well as on low energy cost. In addition, many regions of the United States are becoming limited by transmission capacity and local emission controls. Solar-electric power addresses these issues because it is easily sited at the point of use with no environmental impact. Moreover, because sunlight is widely available, the United States can build a solar-electric infrastructure that is geographically diverse and less vulnerable to international energy politics and volatile markets based on fossil fuels.

The International Energy Agency (IEA) projects that 3000 GW of new capacity will be required globally by 2020, valued at around $3 trillion; IEA also projects that the fastest-growing sources of energy will be supplied by renewables. Much of this new capacity will be installed in developing nations where solar-electric power is already competitive.

Clearly, the nation that can capture a leadership position has the potential for substantial economic returns.

The United States has long been the world's leader in photovoltaic research, technology, manufacturing, and sales. But other countries have awakened to the potential of photovoltaics and its rapidly growing markets. These countries are accelerating their own efforts to secure dominant technologies and global market share. Consequently, over the past few years, the United States has lost its dominant market share and now risks losing its lead in developing and commercializing technology. If we do not rise to the challenge of reestablishing a leadership position, then our domestic PV industry—which includes U.S.-based manufacturers, distributors, and installers—will continue to lose technology leadership, market share, jobs, and revenues. We will be importing PV products to meet domestic demands for electricity—a position similar to the one we are in regarding petroleum.

The U.S.-based PV industry is working hard to meet growing market demand, confront increasing foreign competition, and build a stronger leadership position, because we realize the environmental, economic, and energy security benefits of a large and profitable PV industry within the United States. To do so, we have devised a unified industry roadmap with a vision and long-term strategies, goals, and targets through 2020. We have held two high-level meetings—the first in Chicago and the second in Dallas—in the past year to collectively develop this strategy.

Our roadmap will help to guide U.S. photovoltaic research, technology, manufacturing, applications, market development, and policy through 2020. Its success will depend on the direction, resources, scientific and technological approaches, and continued efforts of the "best and the brightest" among industry, the federal government, research organizations, and our educational institutions.

This roadmap was developed over several years and represents the collective thinking of many experts from industry and government. In particular, three workshops were held to discuss the roadmap activities:

1) NCPV: Workshop on PV Program Strategic Direction, July 14-15, 1997 (Golden, CO)

2) U.S. Photovoltaics Industry PV Technology Roadmap Workshop, June 23-25, 1999 (Chicago, IL)

3) PV Roadmap Conference, December 13-14, 2000 (Dallas, TX)

From these meetings emerged a detailed roadmap that identified the vision for the PV industry, the technical and market barriers, and strategies for overcoming these barriers. The sections that follow provide information on each of these topics.

THE MARKET POTENTIAL FOR PV

Americans are clear about their preferred energy future—plentiful and reliable sources of clean energy at reasonable prices. At the same time, a number of drivers, trends, and issues make meeting these preferences difficult. Global economic and population growth, technological advances, power quality and reliability problems, environmental challenges, and utility restructuring underscore the need and opportunity for re-engineering the nation's energy generation and delivery systems in the coming years. So, although photovoltaics is not the entire solution to these challenges, this renewable-energy option can be an important contributor to the energy picture of the United States and the world.

Photovoltaics has a variety of attributes that will make it an important component of our nation's energy future. PV is a versatile electricity technology that can be used for many applications, from the very small to the very large. It is a modular technology that enables electric generating systems to be built incrementally to match growing demands. PV is easy to install, maintain, and use. It is a convenient technology that can be used anywhere there is sunshine and that can be mounted on almost any surface. PV can also be integrated into building structures to maximize aesthetics and multifunctional value. These positive attributes allow PV to address the following market drivers for energy in the United States.

Reliable Power and Power Quality

The cost of power interruptions is very large, and some customers—for example, those with vital web servers or critical hospital or industrial needs—cannot tolerate power interruptions or poor-quality power. Each year, U.S. businesses spend some $2 billion for industrial uninterruptible power supplies while consumers purchase another 200,000 small generators (about 3 kilowatts) because of concerns regarding power quality and reliability. Furthermore, losses incurred by businesses due to power quality and reliability problems account for more

Figure 9-1. Building Integrated PV System—PV Shingles on Home Roofs

than $30 billion each year.

    Dispersed-generation sources, such as PV, can improve grid reliability by reducing stresses on transmission and distribution systems. Photovoltaic technologies, in particular, provide ultimate power reliability with on-site generation. The reliability of photovoltaics is underscored, for example, by San Francisco's recently announced plan to install photovoltaic-powered traffic stoplights that have backup battery power at 100 key intersections. The City will rely on PV to prevent dangerous traffic snarls during potential rolling blackouts due to a strain on conventional power generation. As noted, in this stand-alone application of PV, a battery backup ensures that power is available even at night or when the sun isn't shining. For a grid-connected application, the grid uses the generated solar electricity "while the sun shines," saving the use of conventional fuel. In effect, this unspent fuel acts as the storage (backup) for electricity needed at other times (e.g., at night).

Plentiful Power Where You Need It

    Solar-electric power systems provide a domestic source of energy that is plentiful, sustainable, and available throughout the United States.

As an example, one utility—Sacramento Municipal Utility District—estimates that sufficient commercial and residential roof space, parking lots, and transmission corridors exist with south-to-west orientation for solar-electric power to provide 15% of its peak power needs. To grasp the potential of solar-generated power, the total electricity demand for the United States today could be supplied by PV systems covering only 0.4% of the nation in a high-sunlight area such as the Desert Southwest—an area about 100 miles by 100 miles. In reality, though, this power generation will be distributed across the United States, bringing generation sources close to the consumer point of use.

Customer Choice

Increasingly, electricity customers want to choose their energy supplier, for both greater control of their power and to illustrate other personal values, such as concern for the environment. Photovoltaic solar-electric power increases customers' choice over the type of energy resource desired. Customers can have PV panels mounted on their homes and businesses. Business and industry owners can generate power for industrial processes and to heat and air-condition multi-family residen-

Figure 9-2. PV Water Pump in Remote Location

tial and commercial buildings and facilities. Businesses can also use PV to meet their critical power needs during power outages and shortages. In a competitive marketplace, customer choice will rely on interconnection and net-metering standards that fairly compensate grid-connected PV users for the energy they generate and that provide safe, secure interconnections with the power grid.

High Value and Appropriate Applications.

Customers—whether residential, commercial, or industrial—recognize that many factors besides price affect the value of energy: for example, power reliability, power quality, freedom from price volatility, or preference for environmentally clean technologies. Utility companies that invest in PV for their customers will be more competitive in the future. As the utility market deregulates, customers will choose their electricity providers based, in part, on the offered price and on desired preferences. As customers learn more about the benefits of PV, solar-electric power will become the power of choice for a larger number of customers and in increasingly diverse applications.

Environmental Quality

Solar-electric photovoltaic systems, which produce no atmospheric emissions or greenhouse gases, make environmental sense for our nation. Compared to fossil-generated electricity, each kilowatt of solar photovoltaics could prevent substantial emissions that endanger our environment and personal health. Typically, on an annual "per kilowatt" basis, PV offsets or saves up to 16 kilograms of nitrous oxides (NO_x), 9 kilograms of sulfurous oxides (SO_x), and 0.6 kilogram of other particulates. In addition, one kilowatt of PV typically offsets between 600 and 2300 kilograms of carbon dioxide (CO) per year. These savings, of course, vary with regional fossil fuel mix and solar insolation.

Building a PV infrastructure provides insurance against the threat of global warming and climate change. For example, a 2.5-kilowatt system covers less than 400 square feet of rooftop area and supplies the necessary electricity for a typical U.S. home. The annual amount of carbon dioxide saved by the system is about equal to that emitted by a typical family car during the same year.

Environmental and power-transmission limitations increasingly constrain our nation's population from increasing or even meeting their power needs. Little room exists to build intra-urban power plants, and public tolerance is low for new transmission lines. As a long-term strategy, distributed PV brings generation to the point of customer use to meet moderate peak loads—electricity when and where it is most needed.

SOLAR ELECTRIC VISION, STRATEGY, AND GOALS

Vision

Vision for Solar Electric Future

Our vision is to provide the electricity consumer with competitive and environmentally friendly energy products and services from a thriving United States based solar-electric power industry.

Our vision recognizes that a sustainable market-based approach will be required to achieve the mission to bring the energy security, environmental, and economic benefits of solar electric power to the United States. To do this we need to provide attractive products from a profitable industry. Coupled with an aggressive market transformation strategy, our goal is to make solar-electric power a significant component of the nation's energy portfolio within the next two decades.

Strategy

We have four strategies for implementing the vision:

1) *Maintain the U.S. industry's worldwide technological leadership.* Technological leadership is necessary both for economic competitiveness and to become a significant contributor to the nation's energy portfolio. Mounting foreign investments have eroded U.S. market share on the business side and have overtaken our R&D lead on the technology front. It is essential to strengthen and expand our investments to secure our future. We must take our core research, development, and other intellectual resources and integrate them with U.S. industry's best interests—resulting in sound and well-conceived programs and investments that clearly support and

guide U.S. PV industry leadership worldwide. A critical element of this effort is sustained partnerships between the U.S. solar-electric industry and national laboratories and universities.

2) *Achieve economic competitiveness with conventional technologies.* During the past 25 years, the cost of photovoltaics has come down by several orders of magnitude. Concurrently, our industry has grown at average annualized rates of 15% to 20%—a growth rate comparable to that of the semiconductor and computer industries. Based on the actual cost of electricity at the point of use, current PV systems are within a factor of 2 to 5 of conventional sources for distributed applications (e.g., residential rooftops). Enormous markets will be established for PV as its cost approaches that of conventional technologies. Our roadmap charts a course that will provide competitive power (i.e., costs of under $3 to $4 per peak watt) in a time frame that will ensure a competitive position.

3) *Maintain a sustained PV market with accompanying production growth.* Sustained growth in production capacity and markets will establish solar electricity as a significant contributor to the nation's energy portfolio, which consisted of about 825 gigawatts (GW) in 2000 of peak electrical generation capacity. Our expectation for industry growth is 25% per year—a level that should be achievable according to recent market data. At this level of growth, domestic PV capacity will approach 10% of U.S. peak generation by 2030. PV will strongly impact AC distributed generation and DC value applications.

4) *Make the PV industry profitable and attractive to investors.* Our aggressive growth strategy will require considerable private investment. Our industry must be profitable and attractive to investors to earn their financial support. To grow into a domestic business with annual revenue of $10 to $15 billion, we must establish strategic guidance and attract foundational funding now.

If our PV industry grows by 25% per year, cumulative installed solar electric systems will grow substantially from 2000 levels of 75 peak megawatts (MW). U.S. generation capacity is projected to grow at 1% to 3% per year; this incremental capacity addition is expected to be about 22 GW in 2020.

We project the following "endpoint" in 2020: "The domestic photo-voltaic industry will provide up to 15% (about 3,200 MW, or 3.2 GW) of U.S. peak electricity generating capacity expected to be required in 2020. Cumulative U.S. PV shipments will be about 36 GW at this time."

Our endpoint is important because we focus not only on a fraction of U.S. electricity generation, but also, on the 15%. PV will shave peak-load demand, when energy is most constrained and expensive. Peak shaving alleviates the need to build new intra-city power plants and transmission lines—projects that burden utility budgets and typically meet with customer resistance. This critical long-term strategy moves the load off the grid and handles peak loads at the point of consumer use—true distributed generation.

If we do not focus on and develop U.S. markets—that is, if the per-centage of U.S. shipments to domestic and international markets remains at the current level—then the 3.2 GW goal in 2020 cannot be met. Without this focus on domestic markets, as well as complementary activities in the global marketplace, the United States' opportunity to serve its citizens and its own national interests will be lost to foreign competition.

Non-domestic markets are significant and will continue to repre-sent a substantial portion of sales—especially in the near-term period of the roadmap. However, the importance of PV technology to the interests of the United States makes it imperative to focus on our domestic mar-kets as major targets for growth, sales, and consumer use.

Goals

We have categorized the goals for the roadmap in two major indus-try target areas: total installed capacity and prices.

Total Installed (Annual) Peak Capacity

This will be at about 7 GW installed worldwide by our domestic PV industry during 2020, of which 3.2 GW will be used in domestic instal-lations. We estimate the mix of applications to be: 1/2 AC distributed generation, 1/3 DC and AC value applications, and 1/6 AC grid (whole-sale) generation. This expectation is based on business plans and market trend projections of the PV industry, and published independent analy-ses. Installed volumes will continue to increase, exceeding 25 GW of domestic photovoltaics during 2030. In 2020, cumulative installed capac-ity in the United States will be about 15 GW, or about 20% of the 70 GW expected cumulative capacity worldwide.

Prices

The system price paid by the end-user (including operating and maintenance costs) will be $3 to $4 per watt AC in 2010. Total manufacturing costs—or the cost to produce the components in the system—are projected to be 50% to 60% of the price of the installed system. The success in 2020 of achieving the vision and these goals will be a hundredfold growth—over 2000 levels—in domestic markets and the U.S. industry. Our roadmap sets the stage for further ramping up of the use of this valuable renewable resource beyond 2020, providing significant portions of U.S. and world electricity generation with an environmentally clean, reliable, and competitive energy source.

BARRIERS: TECHNICAL, MARKET, AND INSTITUTIONAL

Barriers to widespread use of solar electricity reflect technical (e.g., scientific and engineering), market, and institutional problems that can be solved if we as an industry, along with our partners, address them in a unified and complementary manner. We play a key role in removing barriers that block solar-electric technologies from being a prominent power of choice for our nation—a point that became clear during intensive roadmap workshops in Chicago and Dallas involving key PV industry players.

Some solutions where we take a leading role require working with other members of the PV community. Improving intra-industry coordination may even result in formal partnerships among our industry members. Other solutions where we must take the lead will require forming alliances with others, including non-traditional partners. A common theme from the roadmap workshops is that a coalition of forces can bring great power to any potential solution. We are largely responsible for controlling our own destiny.

Technical Barriers

For our industry to reach the goals of this roadmap, we must address a variety of technical issues. One such issue topping the list concerns reducing the cost of manufacturing solar-electric power components. We need to develop low-cost high-throughput manufacturing technologies for high-efficiency thin-film and crystalline-silicon cells. For example, the industry has established an 18% to 20% conversion efficiency goal at a cost of less than 50 cents per watt for each module technology. Currently, thin-film and crystalline-silicon modules are 7% to 10% and 12% to 14% efficient, respectively. In addition, to increase the production of crystalline silicon at the projected rates, a dedicated supply of solar-grade silicon feedstock must be available at less than $20 per kilogram.

In developing our roadmap, we have given considerable attention to the technical barriers facing PV manufacturing processes. These barriers include the need for an improved manufacturing infrastructure to increase throughput and yield. The rate at which PV components are manufactured is still too low, and the projected steady increase in manufacturing output will create even higher demands. Process controls are inadequate, and automation is still insufficient to improve the cost efficiency of production. Continued research and development is needed to improve annual throughputs to about 200 megawatts per factory.

We recommend establishing a Manufacturing Center of Excellence, a key element of which will be an Industry Technology Consortium composed of core equipment manufacturers, PV manufacturing industry representatives, and university/national laboratory research groups. Members would contribute their multidisciplinary expertise for development of programs and facilities for understanding and improving PV component processing and system manufacturing.

The precedent exists for such successful joint research efforts—for example, the Microelectronics and Computer Technology Corp. and SEMATECH (the Semiconductor Manufacturing Research Consortium). An important function of the Center would be regularly held industry forums to develop standards for common equipment and to collaborate on equipment development. These forums would also identify common problems and solutions, including development of standard module electrical and mechanical "interfaces," improved balance-of-systems component reliability, and assistance in developing a more highly trained PV manufacturing labor force.

Another barrier to the widespread use of photovoltaics is the high cost of module materials and encapsulation. In addition, continued R&D on materials and devices must further improve the efficiency of PV systems. Some representative examples of critical R&D needs include high-efficiency thin-film devices, low-temperature interconnect and contact material, and low-cost lattice-matched substrates for compound semiconductors.

We understand that system simplicity and reliability will greatly enhance the widespread acceptance and use of solar electricity, so we aim toward complete systems solutions. As such, a successful PV system should be pre-engineered, pre-packaged, and even "plug-and-play"; highly reliable, long-lived, fault-tolerant, and shade-resistant; and easy to maintain, use standardized components, and sold as a complete service solution.

Successful systems can be achieved in many ways, including the following consensus ideas we formulated during the Chicago and Dallas roadmap workshops:

- Continue to support large numbers of rooftop installations, gleaning valuable systems performance and reliability data for future use

- Educate the PV industry itself on successful systems integration

- Increase the experience of PV systems engineers

- Develop an incentive program—a "Golden Carrot" program—to spur the creation of packaged PV systems

- Continue national meetings on system performance and reliability, jointly sponsored by industry and the national laboratories.

Balance of systems, or BOS, is another area critical to successful PV systems. BOS components include power inverters and other power-conditioning equipment. In the past, less attention was paid to BOS, compared to cell and module manufacturing, when defining our PV industry. Today, however, we clearly realize that we must consider the entire PV installation if we are to achieve our goals. This renewed spirit of collaboration among companies from all segments of our industry is one of the major outcomes of the roadmap process.

Likewise, we in the PV industry are taking the lead in going outside the PV world to develop formal partnerships with inverter manufacturers, to create highly reliable, relatively inexpensive, flexible, trouble-free inverters. In the future, inverters may look more like conventional electronic systems, be free of noise, and incorporate new power electronic topologies.

We realize that inverters must meet qualification tests and satisfy rigid interconnection standards to help stave off the burgeoning influx into the United States of foreign BOS components. Ideally, we will agree on and develop common power-conversion equipment. To address BOS needs, we must initiate a research project representing a collaboration of industry and national laboratories.

Market Barriers

A variety of market-related issues impede the robust development of solar electricity, such as: consumer awareness and education; government, legislative, and regulatory roadblocks; and financing. We understand industry's lead role in modifying marketing strategies. We have identified specific strategies that we ourselves will strive to implement, summarized in Table 9-1.

Figure 9-3. BOS and Inverter Unit for U.S. Postal Service PV System

Table 9-1. Workshop Consensus on Specific Strategies That Industry Must Pursue to Overcome Market Barriers

- Increase value proposition to customers
- Develop alliances with other groups
- Develop a common message
- Form an industry coalition to strategize
- Strengthen the industry's trade association
- Lower product price
- Improve the distribution infrastructure
- Consider developing alliances with energy service companies
- Target end-user groups with appropriate messages
- Reduce all market barriers with a plug-and-play application
- Reduce technical jargon in advertising
- Develop a killer application

Consumers must become better educated about using solar energy—not just for hot water and space heating, but also for their electricity needs. They do not need to worry about understanding the underlying physics of solar-electric generation; but they will want to be firmly convinced of the practicality and performance of PV systems over time. Consumer awareness of and familiarity with solar technologies should start at an early age in educational institutions and should continue into the marketplace. When solar-electric systems become available in home repair and hardware stores, and when consumers are offered installation assistance, PV will then become more "mainstream."

At the same time, the construction, installation, and maintenance infrastructure remains a barrier to widespread use of PV systems, particularly of stand-alone systems. Installation and maintenance professionals must become familiar with solar-electric components and systems so that they can select, install, and maintain them for their customers.

Successfully integrated solar electricity on commercial and residential buildings will significantly boost the marketing of building-integrated photovoltaics (BIPV) by removing a number of "perceived" barriers to its use. For example, we have developed the following key steps to address BIPV. Our short-term actions over a period of 0 to 3 years include striving for architectural integration, and developing wiring systems for curtain-wall applications. Our longer-term actions over a period of 3 to 10 years include: demonstrating examples of good building design and integration; designing value-added building products using PV; and striving for flexibility (e.g., range of colors) in products for architects, designers, and builders.

Other significant market barriers include the need to develop brand-name recognition and pricing for solar-electric components and systems. Currently, consumers—whether residential, commercial, institutional, or government—purchase heating, ventilation, and air-conditioning (HVAC) products (e.g., water heaters, furnaces) by catalog or

Figure 9-4. Building Integrated PV at 4 Times Square in NYC

through vendors who sell specific manufacturers' products. Similarly, the market for PV products will increase through more effective branding and competitive pricing. In sum, the market barriers include:

- Lack of consumer awareness and understanding
- Disincentives against net metering
- Lack of purchasing channels
- Lack of trained installers and inspectors
- Inadequate codes and standards related to PV
- Minimal financing options for PV systems

Institutional Barriers

Institutional barriers remain, including excessive standby and interconnection charges that prohibit integrating PV systems with grid electricity. Even before electricity restructuring spreads across the country, state legislatures and regulatory agencies should be deciding on equitable interconnection charges, standby charges, and net-metering requirements and fees for solar electricity generated in distributed applications and then sold to the grid. Energy customers should find themselves with greater choice under both traditional regulation and retail competition. Where traditional regulation continues, customers ought to be free to pursue more energy efficiency and to acquire distributed generation, including PV. Individuals and organizations who install PV systems must not be "punished" with high charges for interconnection, standby, and sell-back services. Yet, they need to be confident that their distribution utility will work cooperatively with them to allow— and indeed, encourage—grid interconnection. Equity in tax policies for PV compared to other energy sources remains an issue on the state and federal levels.

Institutional barriers to solar electricity development include:

- Lack of communication within industry in identifying common technical problems
- Insufficiently trained and available PV manufacturing labor force
- No solar-electric appliance ratings/standards
- Interconnection standards that inhibit solar-electric development

- Inconsistent government policy related to photovoltaics

The value of photovoltaics is becoming clearer as consumers look to more distributed energy opportunities in our increasingly volatile energy environment. Barriers to a robust PV industry do exist. Nevertheless, the product is basically sound, the market opportunities exist, and our industry's track record has improved dramatically. Although there is much to do, we remain confident that the barriers can be overcome and that cost can be reduced to realize PV's promise. We will pursue the manufacturing of PV products and the use of these commercial products in a broad range of applications within diverse markets. Our industry will also rely on the core R&D activities of the government and universities to help overcome technical barriers and to address the technical issues related to the market and institutional barriers.

Some barriers are best overcome by state or federal initiatives, whereas others are best approached by R&D efforts in academic institutions or national laboratories. Our PV industry members realize that breaking down many other barriers is within their own purview. Continuing to identify and address barriers that are clearly the responsibility of the PV industry will be a critical activity for reaching the goals set forth in this roadmap.

STRATEGIES TO OVERCOME BARRIERS

Photovoltaic solar-electric technology uniquely satisfies the requirements of the three drivers of the new power-generation landscape: premium power for high reliability, distributed generation for point-of-use economics, and renewable energy for environmental value and energy security.

Despite these attributes, the value of this enabling technology is not fully appreciated in the United States. Thus, our industry's commercialization plan will rely on market-driven incentives in federal procurement, tax, deregulation, pollution prevention, and research, development, and deployment. With ever-growing pressures—of energy imports, energy price volatility, power outages, and energy shortages—solar technologies represent a technological safety valve for American home and business owners. We in the solar industry urge that these cutting-edge technologies receive increased attention and that this attention be at least equal to that given by our industrialized competitors in Germany, Japan, and other countries around the globe.

Our PV industry, seeking to address vital energy issues, endorses a roadmap that:

- Tailors research and development programs to address market solutions

- Enhances pollution prevention approaches to focus on clean alternatives

- Ensures customer choice

- Provides targeted tax incentives that seed the market without distorting it

To achieve these goals, we are pursuing the following five strategies:

1) **Develop opportunities based on electric utility deregulation.** Rational deregulation leads to customer choice. Photovoltaic solar-electric power adds unique value in alleviating the problems of supply shortages, price volatility, random and planned power outages, and constraints in transmission and distribution. Required actions involve work in the areas of net metering, consumer education, renewable energy portfolio standards, and system benefits trust funds.

2) **Establish tax equity.** National and state tax incentives, whether investment or production credits or property and sales taxes waivers, must be prioritized for emerging technologies in less mature markets. At the federal level, most energy tax benefits focus on mature energy technologies in mature markets, with estimated federal subsidies ranging from $2 to $8 billion per year. As a whole, renewable energy technologies receive only a small share of these energy subsidies—about $100 million per year of federal tax subsidy—with more than 80 percent of this amount going to wind and geothermal. We recommend the following tax incentives:

   — *Legislatively establish a 15% residential tax credit for solar-thermal and solar-electric installation.* This residential tax credit—as proposed under S. 1634 and H.R. 1465—has been scored by the Joint Tax Committee at $92 million over 5 years. This modest tax credit becomes effective if coupled with a system benefits charge for electricity at the state level.

— *Institute an alternative minimum tax (AMT) waiver similar to that currently enjoyed by domestic oil producers.* Oil producers receive a waiver from the AMT because the national interest is served by sustaining a domestic energy industry, albeit a very small one. The same justification should apply to domestic producers of solar energy.

— *Further expand state incentives.* Currently, 35 states have some type of solar incentive—from investment credits to sales tax and property tax waivers. Such programs help establish tax equity for capital-intensive, fuel-free energy technologies. Also, if adequately promoted to the public, these programs will establish key market-driven incentives for allowing solar technologies to reach a critical market share for sustained growth.

3) **Increase funding of RD&D**. The United States' investment in photovoltaic RD&D has been in the range of $50 to $75 million per year, significantly less than the government's investment in conventional energy technologies. A sufficient baseline investment for federal solar-electric RD&D must be established. In addition, existing programs in other agencies should co-invest in the development and demonstration (deployment) portion of the PV RD&D budget. Specifically, we recommend the following:

— Consistent funding levels of the next 5-year period for photovoltaic RD&D programs through the U.S. Department of Energy of $100 million per year, to maintain technical leadership, which includes the ownership of next-generation technologies.

— Support for validation for PV technology development and deployment by other federal agencies—including the Department of Defense, Environmental Protection Agency, Housing and Urban Development, National Institute of Science and Technology, and other DOE programs.

4) **Establish procurement incentives for federal agencies** The

nation's energy procurement budget should include capital expenditures by federal agencies for cost-effective uses of photovoltaics in four categories: uninterruptible power supplies, lighting, off-grid power systems, and diesel generator replacement.

5) **Target pollution prevention and emissions reduction**. The extraction, conversion, and use of energy is the single largest cause of air and water pollution, as well as of emissions that may lead to global climate change. Solar-electric power technologies are now available that can cost-effectively provide clean, safe, reliable, power.

However, traditional "command and control" regulatory strategies do not promote market-driven approaches to emissions reductions. Newer "allowance trading" programs, whether for clean air or climate change, do not reward the cleanest technologies. Thus, they have not fostered the use of solar and other zero-emission technologies. To this end, we recommend the following initiatives as a fresh approach to solving this problem:

— *RD&D programs, particularly at the Environmental Protection Agency, should provide analytical tools for federal and state environmental regulators and program implementers.* These tools should provide "rules of thumb" for quantifying emissions and pollution-prevention attributes of solar energy—both on a project level and consolidated.

— *Trading programs for clean air and climate-change emissions should reward zero-emissions technologies, rather than least-cost options that provide purely short-term incremental reductions.* The goal for U.S. environmental regulation should be to promote market-driven solutions that translate into the cleanest available technologies and installations.

— *Federal and state promotional and RD&D programs should be leveraged, aggregated, and implemented through states toward technology validation (demonstration) and deployment.* Aggregating the use of solar technologies is the only way to demonstrate and validate significant emissions reductions. To achieve this result, federal and state governments should focus on projects that can be replicated.

— *Government agencies should increase consumer awareness of the cost-*

effective uses of solar technologies. EPA's and DOE's program—which places public service ads and provides logos and consumer awareness—should be broadened for zero-emissions technologies such as solar. National recognition programs, such as that employed by, must also be used to acknowledge early significant users of replicable projects.

— *More aggressive funding should be made available for programs that not only promote solar technologies in schools to reduce energy, but also, as an integral part of the curriculum from elementary through college levels.* According to the Department of Energy, energy is the third highest cost of education after teacher salaries and benefits. Solar energy will offset energy costs and potentially limit increases in property taxes; additionally, though, it will provide future consumers with some "first-hand" experience in clean technologies. Student involvement has made paper and plastic recycling a universal practice over the last 25 years—and such involvement could do the same for solar technologies over the next two decades.

Industry must embark immediately on a more proactive and coordinated program of analysis, market aggregation, consumer awareness and education, and deployment. This effort will significantly reduce pollution through a broad portfolio of solar technologies and applications. Our national goal should go beyond simply cleaning up dirtier fuels and processes. Indeed, our goal should be to enhance the use of the cleanest technologies as a way to drive pollution reduction in a more comprehensive and fundamental way.

CONCLUSIONS

If we are successful in pursuing our overall commercialization strategy, we will create thousands of new, high-value jobs. We will reduce energy imports. We will displace pollution equal to the emissions of one million vehicles. We will provide a more stable energy environment. We will provide energy choice to our citizens. And we will lessen the pressure on energy rates and supply, making sure that the lights will stay on for all Americans.

[**Editor's** Note: This chapter represents an abridged and edited version of the report entitled *Solar Electric Power: The U.S. Photovoltaic Industry Roadmap* published by the Solar Energy Industry Association, May 2001. Much thanks goes to Glenn Hamer, Executive Director of SEIA for permission to publish an abridged version of this report.]

APPENDIX 9-1:
PARTICIPANT LIST FOR WORKSHOP
ON PV PROGRAM STRATEGIC DIRECTION,
JULY 14-15, 1997 (GOLDEN, CO)

Clay Aldrich, Siemens Solar Industries

Tim Anderson, University of Florida

Chuck Backus, Arizona State University East

Allen Barnett, AstroPower, Inc.

Bulent Basol, ISET

John Benner, National Renewable Energy Laboratory

Robert Birkmire, Institute of Energy Conversion, University of Delaware

William Bottenberg, PVI Photovoltaics International Inc.

Chris Cameron, Sandia National Laboratories

David Carlson, BP Solarex

Steve Chalmers, PowerMark

Vikram Dalal, Iowa State University

Michael Eckhart, Management & Financial Services

Alan Fahrenbruch, Stanford University

Todd Foley, BP America Inc.

Christopher Frietas, Trace Engineering

Robert Gay, Siemens Solar Industries

Jessica Glicken, Ecological Planning and Toxicology, Inc.

Ray Gordon, Harvard University

Subhendu Guha, United Solar Systems Corporation

Don Gwinner, National Renewable Energy Laboratory

Brian Huff, The University of Texas at Arlington

Roland Hulstrom, National Renewable Energy Laboratory

Vijay Kapur, ISET

Lawrence Kazmerski, National Renewable Energy Laboratory

Ron Kenedi, Photocomm, Inc.

Edward Kern, Ascension Technology

Richard King, U.S. Department of Energy

Roger Little, Spire Corporation

Rose McKinney-James, Corporation for Solar Technology and Renewable Resources

Hans Meyer, Omnion Power Engineering Corp.

Mohan Misra, ITN Energy Systems

Donald Osborn, Sacramento Municipal Utility District

James Rannels, U.S. Department of Energy

Ajeet Rohatgi, Georgia Institute of Technology

Dan Sandwisch, Solar Cells, Inc.

Richard Schwartz, Purdue University

Mary Shaffner, Solar Energy Industries Association

Jawid Shahryar, Solec International, Inc.

Mike Stern, Utility Power Group

Steven Strong, Solar Design Associates

Tom Surek, National Renewable Energy Laboratory

Margie Tatro, Sandia National Laboratories

Jerry Ventre, Florida Solar Energy Center

Cecile Warner, National Renewable Energy Laboratory

John Wiles, Southwest Technology Development Institute

APPENDIX 9-2:
PARTICIPANT LIST FOR
U.S. PV INDUSTRY TECHNOLOGY ROADMAP
WORKSHOP, JUNE 23-25, 1999 (CHICAGO, IL)

Clay Aldrich, Siemens Solar Industries
Tim Anderson, University of Florida
Allen M. Barnett, AstroPower, Inc.
Bulent Basol, ISET
John Benner, National Renewable Energy
Laboratory
William Bottenberg, PVI Photovoltaics
International Inc.
Gerry Braun, BP Solarex
Jeff Britt, Global Solar Energy
Connie Brooks, Sandia National Laboratories
Chris Cameron, Sandia National
Laboratories
David Carlson, BP Solarex
Steve Chalmers, PowerMark
Clint (Jito) Coleman, Northern Power
Systems
Maurice Covino, Spire Corporation
Ghazi Darkazalli, GT Solar Technologies,
Inc.
Alan E. Delahoy, Energy Photovoltaics,
Inc.
Tom Dinwoodie, PowerLight Corporation
Erten Eser, Institute of Energy Conversion, University of Delaware
Jim Galica, STR
James Gee, Sandia National Laboratories
Subhendu Guha, United Solar Systems
Corporation
Jack Hanoka, Evergreen Solar, Inc.
Roland Hulstrom, National Renewable
Energy Laboratory

Joe Iannucci, Distributed Utility Associates
Masat Izu, Energy Conversion Devices,
Inc.
Theresa Jester, Siemens Solar Industries
Robert Johnson, Strategies Unlimited
Juris Kalejs, ASE Americas
Lawrence Kazmerski, National Renewable Energy Laboratory
Richard King, U.S. Department of Energy
David Lillington, Spectrolab, Inc.
Hans Meyer, Omnion Power Engineering
Corp.
James Rand, AstroPower
Ajeet Rohatgi, Georgia Institute of
Technology
Bill Roppenecker, Trace Engineering
Bob Shaw, Arete Ventures, Inc.
Chris Sherring, PVI Photovoltaics
International Inc.
Mike Stern, UPG Golden Genesis
Tom Surek, National Renewable Energy
Laboratory
Jerry Ventre, Florida Solar Energy Center
Howard Wenger, AstroPower, Inc.
Chuck Whitaker, Endecon/PVUSA
John Wiles, Southwest Technology
Development Institute
Paul Wormser, Solar Design Associates
Inc.
Jan Brinch, Melissa Eichner, Robyn
McGuckin, Joseph Philip, Kim
Reichart, Jennifer Ryan, Richard Scheer,
Paula Taylor, Energetics, Incorporated

APPENDIX 9-3:
PARTICIPANT LIST FOR
PV ROADMAP CONFERENCE,
DECEMBER 13-14, 2000 (DALLAS, TX)

Rajeewa Arya, BP Solar

John Benner, National Renewable Energy Laboratory

Bob Birkmire, Institute of Energy Conversion, University of Delaware

Gerry Braun, BP Solar

Jan Brinch, Energetics, Inc.

Connie Brooks, Sandia National Laboratories

Steve Chalmers, Power Mark

Jerry Culik, AstroPower

Alan Delahoy, Energy Photovoltaics

Jennifer Dunleavey, Energetics, Inc.

Jim Dunlop, Florida Solar Energy Center

Chris Eberspacher, Unisun

Andrew Gabor, Evergreen Solar

James Gee, Sandia National Laboratories

Christy Herig, National Renewable Energy Laboratory

Tom Huber, S&C Electric

Roland Hulstrom, National Renewable Energy Laboratory

Terry Jester, Siemens Solar Industries

Juris Kalejs, ASE Americas

Larry Kazmerski, National Renewable Energy Laboratory

Edward Kern, Applied Power

Richard King, U.S. Department of Energy

Paul Klimas, Sandia National Laboratories

Dave Lillington, Spectrolab

Paul Maycock, PV Energy Systems

Ron Pitt, Xantrex Technology

Ajeet Rohatgi, Georgia Institute of Technology

Rich Scheer, Energetics, Inc.

Pete Sheldon, National Renewable Energy Laboratory

Alison Silverstein, Public Utility Commission of Texas

Ed Skolnik, Energetics, Inc.

Tom Surek, National Renewable Energy Laboratory

Blair Swezey, National Renewable Energy Laboratory

Joe Tillerson, Sandia National Laboratories

Bob Walters, ENTECH

Chuck Whitaker, Endecon Engineering

John Wiles, Southwest Technology Development Institute

Paul Wormser, Solar Design Associates

Bob Yorgensen, Specialized Technology Resources

Chapter 10

Conclusion—
Learning from Roadmaps

INTRODUCTION

The roadmaps contained in this book provide a rich array of information on sustainable energy technologies and their promising contributions to our future energy landscape. The questions addressed in this chapter are: (1) What can we learn from these roadmaps regarding the usefulness of "roadmapping" in the strategic planning process? and (2) What can we learn about our energy future by reviewing these roadmaps collectively? We explore these questions by first evaluating the similarities and differences of the roadmaps, and then assessing roadmap outcomes in light of existing forecasts of energy production and consumption to year 2025.

Similarities and Differences in Roadmap Process and Content

There are a number of similarities and differences that emerge from the roadmaps presented in this book. Exploring these similarities and differences can help us draw some conclusions about the usefulness of roadmaps as a planning and policy tool.

Similarities

The roadmaps reviewed herein had much in common, as summarized in Table 10-1. Perhaps the most striking similarity is in their "tone." All of the roadmaps are optimistic about the opportunities for sustainable energy technology deployment in the future. Although in all cases obstacles were identified, none was presented as insurmountable—each roadmap argued that with the right mix of technology, policy, and market forces these sustainable energy technologies could find a solid foothold in future energy sectors. The transition from conventional (i.e., fossil) fuels and power production to sustainable alternatives is presented as something that could be accomplished over the next 20-25 years.

Table 10-1. Similarities among the roadmaps

Similarities

- Optimistic tone on the future of sustainable energy technologies
- Market transformation issues are as important as technology development issues and market incentives are required to "jump start" the market
- Focus on public-private partnerships and industry collaboration
- Roadmap considered to be "living document"
- Role of government to collect and disseminate information and to coordinate codes and standards for technology development
- Vision statements include identifiable goals and time frames
- Input from a broad array of stakeholders

However, this optimism should be tempered with historic reality. One need only look back 25 years to find numerous predictions for a year 2000 "future" relying on solar power, hydrogen, and electric vehicles. Alas, this future has not yet arrived—the barriers being too significant and the markets too heavily weighted towards fossil fuel production and use. The question now is whether the strategies identified in the roadmaps presented here are sufficient to overcome these barriers in the next 25 years.

All of the roadmaps identify *market transformation* as critical to successful technology deployment. This transformation will occur through improved financing, life cycle costing, environmental accounting, standards and codes development, and increasing public awareness. Often, these kinds of transformations trump the technological transformations, especially for roadmaps that have mature technologies that simply need to develop market niches (for example in the buildings and lighting roadmaps).

The emphasis on market transformation leads to a strategic emphasis on incentives to push or pull markets for sustainable energy technologies. Many of the roadmaps exhibit an "if-only" flavor (i.e., "…if only the proper incentives were in place, our technology would see significant market penetration"). The "if-only" sentiment is understandable; the developers of these technologies truly believe that implementing sustainable energy systems is technically feasible, environmentally necessary, and political wise. Thus, market imperfections are holding back what ultimately are sensible energy decisions. Incentives that reduce these imperfections make good policy sense. However, how far is government expected to go to support these technologies? The answer to this question may dictate the ultimate success or failure of sustainable energy technologies.

Another common theme in the roadmaps is the focus on public-private partnerships and industry collaboration to develop and promote sustainable energy technologies. The roadmaps emphasize public-private RD&D projects, particularly since many of the technologies pose high-risk premiums for investors (and thus research is typically under-funded by the private sector alone). In addition, because most of the roadmaps highlight the importance of integrated sustainable energy "systems" (see the *hydrogen roadmap* or the *commercial buildings roadmap* for explicit examples), industry partners must work together so that each part of the "system" fits together well. To take the hydrogen roadmap as an example, new distribution and storage technology development for hydrogen must be coordinated with end-use technologies so that consumers see a production-distribution-storage-product system that provides reliable energy services cost-effectively.

Since many of these technologies require public-private collaboration, it should be reiterated that one of the most important outcomes from the roadmapping process is the networking opportunity it grants private and public organizations. Through the roadmap development process, technology development problems are discussed and solutions explored with input from a variety of stakeholders. Some of these stakeholders will ultimately form partnerships that can help advance their particular technologies.

Another similarity was the recognition that these roadmaps are living documents. They represent strategic plans that will be revisited and revised as market and technology conditions change. (Indeed, the solar power roadmap represents a revision of a similar roadmap conducted several years earlier). Some of the roadmaps explicitly identified regular visitation as an important activity, and one suspects that roadmap revisions could be performed every five years or so, assuming resources and a coordinating body still exist to take on this task.

The roadmaps also had a similar view of the use of government as a source for collecting and disseminating information on technologies. In some cases, this involved public awareness and education campaigns; in others it involved the development and dissemination of codes and standards. The role of information disseminator is a common and accepted role for government and is supported by private companies who believe that the public trusts information from government sources more so than information from product vendors.

Each of the roadmaps had a vision statement in some form or another. These vision statements varied in style and content, but an interesting similarity was in the identification of goals and time frames for the technology. For example, some roadmaps identified a percentage of market share that would achieved by a given time frame; others identified growth rates over time; still others highlighted overall production capacity (e.g., measured in MW) over time. The formulation of the vision statement is important, as it provides stakeholders an opportunity to frame the debate over technological development. Often, creating the vision statement is the most difficult part of the roadmapping process.

A final similarity involves the roadmap process itself. All the roadmaps discussed in this book involve significant contributions from a diverse group of stakeholders. Each roadmap provided some mechanism for stakeholder input, either through conferences, committees, surveys, or voting mechanisms. This open process allowed for a diverse range of perspectives ultimately makes the roadmaps richer documents.

Differences

Although similar in many ways, the roadmaps had two obvious differences. These are summarized in Table 10-2 and are discussed in more detail below.

Table 10-2. Differences in Roadmaps

Differences
• Type of organization coordinating Roadmap activity
• Range of technology specificity

One important difference rests with the type or organization that was tasked with coordinating the roadmap. The first four roadmaps in this collection (lighting, commercial buildings, residential buildings, and windows) were coordinated by government organizations, namely the U.S. Department of Energy. As such, these government-coordinated roadmaps stress the role of government as a vehicle for technical R&D and information dissemination. Other roadmaps (hydrogen, biomass) had quasi-governmental coordination (a mix of government and private organizations leading the effort), or had direct private sector coordination (solar, wind). The roadmaps with more private sector coordination stressed public-private partnerships, typically with an emphasis on publicly supported RD&D. The privately coordinated roadmaps also stressed financial incentives (such as tax breaks for sustainable energy technologies) as important public policy options.

Another important difference was the level of specificity that the roadmaps placed on technology development. Some of the roadmaps were very specific about the types of technologies that needed to be pursued. For example, in the residential building roadmap and the windows roadmap, participants were very specific in the types of technologies that needed to be developed. However, in other roadmaps, only technology areas were identified. For example, in the hydrogen roadmap it was mentioned that new forms of hydrogen delivery needed to be developed. These differences clearly arise due to the current state of technological development for each technology. Those technologies that are still in the nascent stages of development are still involved in basic R&D and have less specific development agendas than technologies that are looking for a specific technological refinement on an existing product.

Lessons Learned: The Roadmapping Process

Based on the roadmaps, we can derive some lessons learned for conducting roadmapping activities. These are:

- Important to get diverse people involved and give them a voice;

- Important to have a neutral party that will coordinate the process, so as not to bias the process—credible results require unbiased assessment. Note that a neutral party may be a private sector entity—what is more important than public/private affiliation is the manner in which the roadmapping process is conducted;

- Important to recognize the stage of development of the technology and to adjust expected results based on that stage of development; for example, if a roadmap is being conducted for an immature technology, stakeholders may expect more general roadmap results;

- Important to look at the roadmap as a living document and plan to revisit this document every few years;

- Important to maintain the networking that arises from roadmap development, either through a listserv, conferences, newsletters, or regular meetings;

- Dates and quantified goals are not necessary, but they do provide concreteness to an otherwise abstract concept. (The other advantage of discussing dates and quantified goals is procedural; this activity helps frame the debate among participants leading to more concrete discussions of technological status.)

ROADMAPS AND OUR ENERGY FUTURE

Energy Trends in the U.S.

From these roadmaps, perhaps a clearer picture of our energy future can be discerned. Collectively, these roadmaps highlight a number of important areas for sustainable energy production and use in the future. What can we learn about the future from this collection of information?

To place this discussion in proper context, it is perhaps worthwhile to consider forecasts of our energy future out to 2025. For the United States, the Energy Information Administration conducts such long-range forecasts annually (*Annual Energy Outlook 2003 with Projections to 2025*).[1]

First, consider Figure 10-1, which shows the expectations for increasing energy consumption in the United States to 2025. As shown in this figure, energy consumption is increasing at a faster rate than population—thus implying that per capita energy consumption is also increasing.

Figure 10-2 helps clarify where this increase in energy consumption is derived. Figure 10-2, demonstrates that energy intensity is expected to decrease [measured in 1000 Btu per dollar GDP (1996$)]. We are becoming a more energy efficient economy that is continuing a transition to a primarily service-based economy. Despite improvements in our energy efficiency, total energy consumption still increases (see Figure 10-1) due to increased economic growth and population.

All economic sectors will contribute to this increased growth (as shown in Figure 10-3), with transportation and industry representing about 60% of total energy consumption. Commercial and residential energy consumption will constitute the remaining 40%. It should be noted that transportation demonstrates the largest increase over the next 25 years, yet represents the sector in which the U.S. is most dependent on foreign fuel.

The type of fuel we will consume in 2025 is expected to still be dominated by fossil fuels, as shown in Figure 10-4. Here, over 40% of our total energy consumption will be from petroleum, followed by natural gas (26%) and coal (21%). Indeed, only about 6% of all energy production is expected to come from renewable resources, and most of that is from hydroelectric power production.

Because our energy future is one based (primarily) on fossil fuels,

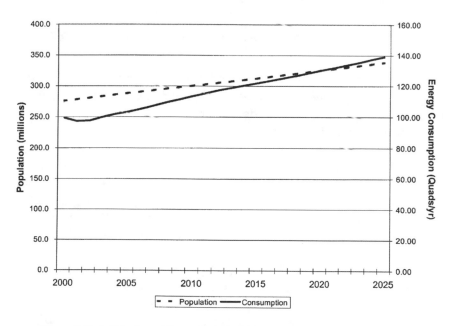

Source: Table 1, EIA, *Annual Energy Outlook,* 2003.

Figure 10-1. Total Energy Consumption and Population to 2025

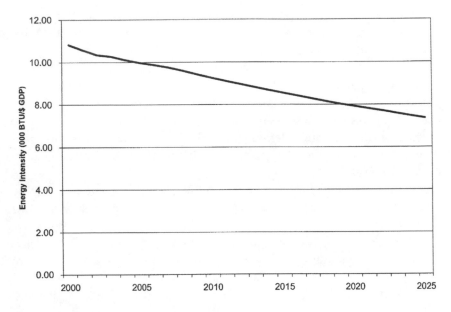

Source: Table 20, EIA, *Annual Energy Outlook*, 2003.

Figure 10-2. Energy Intensity for the U.S. Economy to 2025

the dangers of supply disruptions and resource depletion (discussed in Chapter 1) will seemingly continue to haunt the U.S. This is especially true for petroleum. Figure 10-5 demonstrates that crude oil production in the U.S. will continue to decline or stay flat over the next 20 years. Thus, increases in petroleum consumption will come from imports, as shown by the "Petroleum Gap" in the figure. This gap will be about 13 million barrels of oil per day by 2025. Because the transportation sector in the U.S. is about 97% reliant on petroleum (and is expected to continue to rely on petroleum for its energy needs in the foreseeable future), this dependence continues to cause political and economic concerns for the United States.

Although the roadmaps indicate promise for a renewable energy future in 2025 (particularly biomass, wind, and solar), the U.S. DOE is less than sanguine. For example, in the electricity sector, where renewable energy in the form of solar photovoltaics, wind, biomass, and other technologies hold promise, DOE forecasts that renewable power will still represent only a small slice of the energy picture, as shown in Figure 10-6. Coal will still be the dominant feedstock for electricity production,

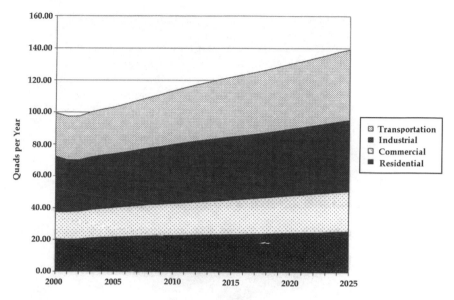

Source: Table 2, EIA, *Annual Energy Outlook*, 2003.

Figure 10-3. Energy Consumption by Sector to 2025

followed by natural gas, and then nuclear power.

Of the renewable power shown in Figure 10-6, almost 70% is from conventional hydroelectric production. Figure 10-7 presents a breakdown on projected renewable electricity capacity for the year 2025. We see from this figure that wind power has made some significant strides in renewable electricity markets; however, this is still a small contributor to total electricity generation.

As mentioned in Chapter 1, in addition to resource issues, environmental issues are driving the interest in sustainable energy technologies. In particular, carbon dioxide and other greenhouse gas emissions has become the environmental problem of greatest concern in the 21st century. Yet, despite warnings of significant climate change and its costly impacts on the U.S. and others, the U.S. is expected to continue to increase carbon dioxide emissions over the next 20 years, topping 2,000 million metric tons of carbon equivalent around 2017 (see Figure 10-8). In addition, Figure 10-9 demonstrates that per capita carbon emissions are also expected to increase, thus total carbon emissions increases are not only due to population growth, but also to increasing use of fossil fuels in the U.S. economy.

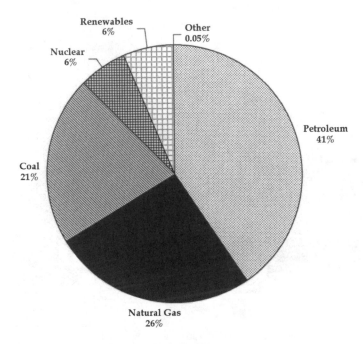

Source: Table 1, EIA, *Annual Energy Outlook*, 2003.

Figure 10-4. Energy Consumption by Fuel Type for the U.S. in 2025

Roadmap Trends and Promise

The projected energy trends presented above are certainly subject to change due to technological breakthroughs and economic incentives—and all the roadmaps share optimism that sustainable energy technologies can find market opportunities over the next 20 years that will lead to a sustainable energy future. Despite what DOE forecasts predict, experts in the renewable energy and energy efficiency fields continue to argue that the time has come for a rapid transition to advanced and sustainable technologies.

Recognizing that such a transition is a Herculean task, these experts have made the case that coordinated strategic planning is *absolutely necessary*. Whatever our energy future holds, they argue, some coordinating body must continue to develop strategic roadmaps that will give direction to industry and government activities.

Indeed, I argue that a more broad-based sustainable energy roadmap initiative is required that should include all sustainable energy

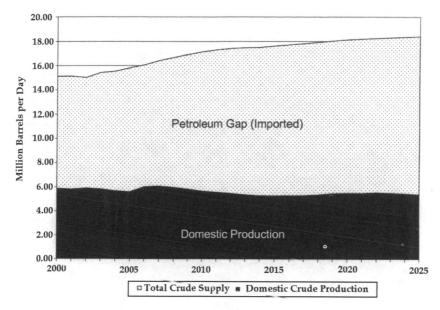

Source: Table 11, EIA, *Annual Energy Outlook*, 2003.

Figure 10-5. Crude Oil Consumption and Domestic Production to 2025

technologies in a more holistic fashion. Perhaps the wind, solar, hydrogen, biomass, lighting, windows, and buildings stakeholders would gain much by working together on a sustainable energy roadmap initiative that allows them to share information, clarify research needs, network, and identify policy issues of mutual interest and support.

There is surely a *need* to approach our energy future with a systems perspective. In many of the roadmaps, technologies overlapped; for example, a future commercial building that includes efficient lighting, advanced windows, and integrated solar PV requires knowledge from four of our roadmap areas. Therefore, it is argued here that a coordinating body (such as the U.S. Department of Energy) should continue to conduct individual roadmap activities, but also be cognizant of overlap and how advancements in one energy field may present important opportunities (or barriers) for another energy field. In addition, DOE should support a roadmap Office that regularly coordinates a Sustainable Energy Roadmap Initiative using input from many different sustainable energy fields.

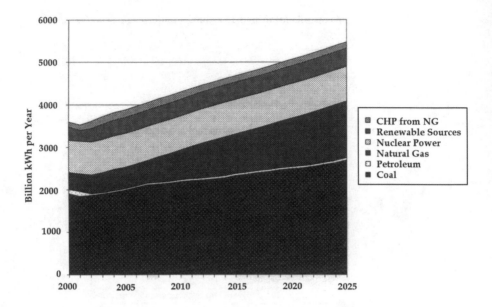

Source: Table 8, EIA, *Annual Energy Outlook*, 2003.

Figure 10-6. Electricity Generation by Fuel to 2025

CONCLUSION

Despite the enormous barriers that some sustainable energy technologies may have to overcome to compete against conventional technologies, I still remain optimistic. With an appropriate mix of private sector ingenuity and public sector incentives and support, we may find sustainable energy technologies working into mainstream America and abroad sooner rather than later.

Through the development of these technology roadmaps, organizations have learned much about the important issues facing their respective industries. They have also begun a process of networking and strategic planning with a common vision in mind. This, certainly, is the right first step if ever these technologies have a chance.

Reference
[1]Energy Information Administration, *Annual Energy Outlook 2003 with Projections to 2025*, DOE/EIA-0383(2003), January 2003.

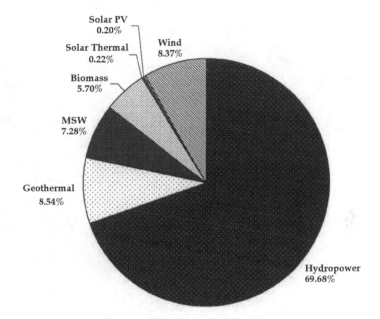

Source: Table 17, EIA, *Annual Energy Outlook*, 2003.
Note: MSW = Municipal Solid Waste

Figure 10-7. Projected Renewable Electricity Capacity by Fuel in 2025

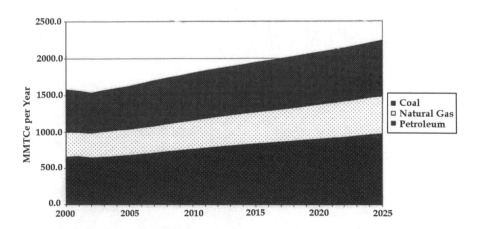

Source: Table 19, EIA, *Annual Energy Outlook*, 2003.

Figure 10-8. U.S. Carbon Emissions to 2025 by Fossil Fuel Type

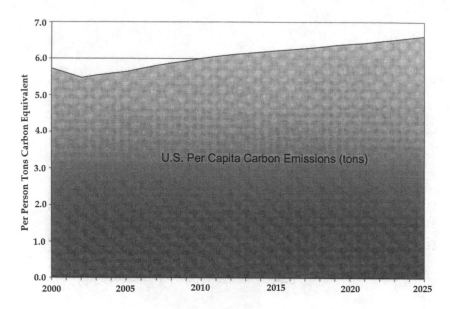

Source: Table 19, EIA, *Annual Energy Outlook*, 2003.

Figure 10-9. Per Capita Carbon Emissions to 2025

Appendix A

List of Photo Credits

Pictures in this book are publicly available and taken from the National Renewable Energy Lab (NREL) Photographic Information Exchange (www.nrel.gov). These photos were not part of the original roadmap documents used in this book. Photo credits are found in the table below. All other figures were generated by the author.

Figure 2-1   CREE Lighting
Figure 2-2   Robb Williamson
Figure 2-4   Lawrence Berkeley National Lab
Figure 3-1   Robb Williamson
Figure 3-2   Warren Gretz
Figure 4-1   Oak Ridge National Lab
Figure 4-2   Enermodal Engineering LTD.
Figure 5-1   Warren Gretz
Figure 5-2   Lawrence Berkeley National Lab
Figure 5-3   Atlantis Energy, Inc.
Figure 6-2   Warren Gretz
Figure 6-3   Mike Linenberger
Figure 6-4   Kevin Chandler
Figure 6-5   SunLine Transit Agency
Figure 6-6   Matt Stiveson
Figure 7-1   Warren Gretz
Figure 7-2   Warren Gretz
Figure 7-3   Warren Gretz
Figure 8-1   Bergey Windpower Co., Inc.
Figure 8-2   Elliott Bayly
Figure 8-3   David Parsons
Figure 9-1   Solar Works
Figure 9-2   Jerry Anderson
Figure 9-3   U. S. Postal Service
Figure 9-4   Andrew Gordon Photography and Fox & Fowle Architects

Index